Decotam

Ball Pest & Disease
M A N U A L

Charles C. Powell, Ph.D.
Richard K. Lindquist, Ph.D.

Ball Publishing
Geneva, Illinois USA

Ball Publishing

335 N. River St., P.O. Box 9
Batavia, Illinois 60510 USA

98 97 96 95 94 93   10 9 8 7 6 5 4 3 2

Geo. J. Ball Inc. hopes that users of this book will find it useful and informative. While the authors have endeavored to provide accurate information, Ball asks users to call its attention to any errors. The authors have attempted to obtain information included in this book from reliable sources; however, the accuracy and completeness of this book and any opinion based thereon are not guaranteed.

Library of Congress Cataloging in Publication Data

Powell, Charles C.
   Ball pest and disease manual /Charles C. Powell, Richard K. Lindquist.
      p.   cm.
   Includes index.
   ISBN 0-9626796-4-X (hardcover) :
   1. Plants, Ornamental—Diseases and pests—Handbooks, manuals,
etc. 2. Plants, Ornamental—Diseases and pests—Control—Handbooks,
manuals, etc. 3. Pesticides—Handbooks, manuals, etc.
I. Lindquist, Richard Kenneth, 1942-   . II. Title.

SB608.07P69   1992
635.9'2—dc20   91-35855
                                                              CIP

Cover photos: Powdery mildew infections can occur on flower tissue of many plants, such as on this gerbera. Notice how easily the superficial fungal growth can be seen when the light is angled onto the petals (top). Spores from rust lesions are most commonly seen on lower leaf surfaces (bottom).

Contents

Foreword

*A*fter reviewing the *Ball Pest and Disease Manual*, I have been excited about its contribution to plant production. My job puts me in contact with growers looking for practical solutions to everyday problems. The authors also work with growers on production-problem solving, perform research and teach in their respective areas. They have authored many technical papers, bulletins and tips on disease and insect control. Dr. Lindquist and Dr. Powell emphasize that plant health is the key to reducing insect and disease problems. In plant production the grower must minimize exaggerated plant stress to produce an economic crop for a specific market. Growers are also responsible for reducing environmental risk by minimizing chemical use for insect and disease control.

The *Ball Pest and Disease Manual* discusses insects and diseases detection and control strategies to help growers decide which biological, environmental or chemical measures can control the problems without damaging the crop. With the combination of western flower thrips and tomato spotted wilt virus we have seen that insect and disease control requires an integrated pest management (IPM) strategy. An IPM program helps the grower produce a healthy, virus-free crop for the consumer. Our first line of defense is to keep the insect or disease out of our production facilities. The second is to detect problems before they are epidemic. The third is to eradicate the problem.

Not only is there specific information on diagnosing problems, there are also several handy reference guides for plant production. Having this information will reduce growers' time locating specific information on producing healthy crops. Reducing production time loss, plant loss, quality loss and labor involved because of insect and disease problems decreases production cost. Considering preventive measures, environmental controls including fans and screens, labor, chemicals and predators, and scouting and monitoring, the production cost might amount to as much as 15 percent of total production costs.

The *Ball Pest and Disease Manual* is a hands-on production tool for plant health management. Everyone responsible for plant production will benefit from having this handy reference for insect and disease identification and control.

> Ron Adams
> Technical Services Manager
> Ball Seed Co.

Chapter 1

Flower and Foliage Crop Health Management:

An Integrated Approach

*P*lant health maintenance is your primary objective as a flower and foliage crop grower in the greenhouse or outdoors. People reach this objective by understanding the links between problems of environmental change or stress and plant processes. This forms the basis of integrated plant health management.

The approach isn't new. It's rooted in the actions of experienced growers who are guided by common sense. Scientists and researchers recently have begun work in this area—work that is often called biological pest control or environmental stress management. This chapter will discuss some of these integrated ideas and the ways they fit together. By using these concepts, you can make total plant care approaches work for you. You will become holistic plant health managers!

The plant health balance

What's a healthy plant? Is it the same as a plant in good health? Not necessarily! A plant may be healthy today but not in continuing good health for tomorrow. Good health involves a balance. If all environmental elements influencing the plant are within reasonable ranges—both by themselves and in relation to each other—the result is a plant that can balance its internal processes to satisfy its needs. It's a plant in continuing good health.

Influential environmental elements always relate to one another. Internal plant processes also always relate to one another. This makes the idea of plant health integrated and holistic.

The basic environmental elements that promote plant health are simple: a friable soil, a proper nutrient balance, sufficient balance of soil acidity, enough root and crown space, ample water, optimum temperature and light, pure air and freedom from pests and diseases. When one or more of these elements is out of range, environmentally induced stress results, placing the plant's health balance at risk.

1

Growers who understand plant health management, such as Claude Hope of Linda Vista Farms in Costa Rica, realize it's an integrated process.

Flower and foliage crops growing in a field or in a greenhouse have remarkable abilities to balance their internal processes and maintain health. They must—they cannot move to escape an inhospitable environment like humans and most other animals can. As plant health managers, we must plan and conduct activities that enable the plants to balance their internal processes. This keeps them healthy.

In many cases, environmental elements leading to plant stress aren't defined precisely when dealing with plant health. But this doesn't make stress less important. Remember that a healthful environment for plants contains many elements that occur in ranges or degrees. These elements include either too much or too little light, temperature and water, and can cause stress.

Acute versus chronic stress

Stresses can be different types, including acute and chronic stresses. Acute stresses occur suddenly and cause damage soon. Improper sprays, frosts or freezes, and injuries from farm implements are acute stress causes. Chronic stresses, on the other hand, can be caused by nutritional imbalances, weed competition or improper soil pH (leading to nutrient unavailability). Chronic stresses take time to work on a plant to make it unhealthy.

Dealing with chronic stresses sometimes is easier than dealing with acute stresses. You have some time to reverse the imbalance soon after stress begins if you learn to recognize signs of chronic stress. An acute stress, on the other hand, gives little time to correct it. About all you can do is learn from the experience and make sure it isn't repeated.

Understanding plant decline

Chronic stresses eventually cause sick plants. The time it takes to make the plant sick is called a period of plant decline. There are all degrees of plant health, from magnificently healthy to pathetically diseased. The longer a plant

These plants, under stress from poor drainage, suffer from root rot.

has to endure a stress, the more it slips along the continuum from health to disease. The key to stress management is to recognize early when decline is beginning and take quick measures to reverse it.

Good environmental awareness can help you recognize when plant stress is apt to occur. When confronted with an unhealthy plant, your first job is to identify chronic, stress-inducing situations even though you already might have noted a particular acute stress, such as an infectious disease or insect pest. Put yourself in the plant's place. Investigate the soil water level, light or temperature extremes, or the soil aeration and compaction. They may be additive in their effects. For instance, soil dryness doesn't become stressful until temperatures climb. Such situations are called disease or stress complexes.

Noninfectious and infectious diseases and pests

"Disease" is the term used to refer to plants showing obvious, imbalanced health symptoms. Sometimes these symptoms result from acute stresses such

Acute stress on poinsettia injured by pesticide spray.

When confronted with an unhealthy plant, try to identify chronic, stress-inducing situations.

as those caused by pesticide burns. In the past symptoms caused by acute stresses have been called injuries. Now some authors are beginning to use the term "disease" even for this plant problem. Symptoms caused by chronic stresses are often called disorders or noninfectious diseases.

Pathogenic organisms and many insect pests commonly attack and infect a stressed plant that already shows symptoms of a noninfectious disease con-

Pathogenic microbes often infect plants already under chronic stress. Mums growing in a wet spot are attacked by Rhizoctonia.

First, diagnose probable problem causes; then make environmental management decisions to reduce stress.

dition. When the pathogen or pest is involved, the health imbalance and stress of the infectious problem is added to the preexistent noninfectious disease.

In such cases, remember that two types of disease imbalance exist in the plant or crop simultaneously. Stress management can help management of infectious as well as noninfectious disease problems. Although some pests and disease organisms attack even vigorously growing and reasonably healthy plants, damage is usually minimal.

It's difficult to provide general guides for stress management in our diverse industry of growing flowers and foliage plants. For the most part, following the growing principles given later in this book will provide stress management. If you do get into a problem, asking the following questions should help you make proper stress management decisions. These questions form the basis of diagnosis, discussed in the next chapter.

1. What kind of plant are you dealing with? What are the plant's cultural requirements? Are these requirements being met? Knowing the plant type will enable you to anticipate what specific diseases or insect pests to expect and when to expect them.
2. What's wrong? Exactly what symptoms are of concern? What is the condition of surrounding crops? Make a list of all the symptoms you see.
3. How long has the problem been going on? Be careful here! Many people sincerely believe plants get sick overnight. This situation is rare, except following acute stresses.
4. What are the probable causes? List causes, including environmental conditions and anything else you think might be contributing to the situation. If necessary, have a plant disease clinic assist you.
5. What are the possible remedies or health management practices? Next to the causes, list all possible remedies, regardless of cost or practicality. Don't forget to include throwing the crop out!

6. Of the possible remedies, which are practical? This is a crucial step! Remember, doing nothing and merely tolerating the situation may turn up as the "most practical" thing to do.
7. For the practical remedies, when would be the best time of year to do them? Planning to avoid problems in the next crop is crucial.
8. Are you truly willing to commit yourself to the time, labor and expense of managing the problem?

As you read the following chapters on insect and disease control, note references to stresses that may be contributing to the problem. Stress management is basic to healthy flower and foliage crop production.

Chapter 2

Diagnosing Problems

Being a good flower or foliage crop production specialist involves being able to quickly identify the problem causes. This is the business of diagnosis.

Diagnosis consists of three general areas. First, perception of a problem; second, determination of the cause or causes; and third, planning a solution to the problem. An important part of the diagnostic procedure is to proceed through these three general areas with organized thinking.

Perception of problems

Perception of problems generally starts with the appearance of a symptom. A symptom is defined as a noticeable, abnormal condition. A good diagnostician recognizes a symptom before the untrained eye can see any symptoms at all, but a good diagnostician doesn't want to overreact to minor situations.

Consider four perspectives to increase your ability to perceive a symptom. One is the detailed or close-up view. This most common way people look for symptoms can be very rewarding. You can use a hand lens magnifier to view plant tissue and quickly diagnose problems such as spider mites or powdery mildews.

In many cases, however, the general view is just as important to notice as the close-up. This general view is the second of the four perspectives. For example, a general view might enable you to determine the source of mites or mildew. Another useful perspective is to get some idea of how long a certain symptom or set of symptoms has been present. Plants cannot tell you how long they have been sick, but you can learn to evaluate and recognize the time involved by using indirect methods and knowledge. For instance, experienced growers know how to keep plant condition records routinely so they can trace back through them when problems arise. Perhaps a lower leaf browning appears on a geranium crop in April. A review of irrigation procedures may reveal that a new watering procedure or fertilization program began in February. This would

A close-up diagnostic view of downy mildew on snapdragon.

suggest that the leaf condition may relate to poor root health initiated by stresses associated with the February changes.

Finally, remember the perspective your knowledge and experience can give you regarding symptomatology. As we study more about plants, we learn to look for certain problems in particular situations or on particular plants. Scheffleras, for example, are especially susceptible to mites. Grape leaf ivy often has powdery mildew. Geraniums get Botrytis diseases easily. An appendix at the end of this book will help you learn what problems to expect in various plants.

Determining problem causes

Generally, plant problems don't arise from a single cause. There may be a primary or most obvious cause such as the occurrence of spider mites or powdery mildew fungus disease, but some associated environmental stress conditions may be present and need diagnosis. You can sort out causes by

Enlarged view of two-spotted spider mite and egg on leaf underside.

A general diagnostic view shows that rose powdery mildew is more severe on the shaded bench to the left.

correct perspective and orderly thinking. Formulate a series of questions to ask about the plant material. You may need to write out lists of symptoms and questions. The key is order. Do not repeat yourself or go off on a tangent unnecessarily.

The loss of leaves on the schefflera does not specifically indicate a cause. Spider mites, the cause of leaf drop here, can be be seen with a closeup inspection.
Photo courtesy of National Foliage Foundation.

One of the most difficult diagnostic areas for the production manager is determining the causes of nonspecific symptoms. Although nonspecific symptoms such as leaf yellowing, leaf drop or leaf edge browning may be serious and easily detected, they don't necessarily relate to a specific cause. Hundreds of environmental problems may result in such symptoms. Some may be infectious diseases and others may be noninfectious, environmental problems. The table included at the end of this chapter gives some examples of nonspecific symptoms and possible causes.

The only way to correctly determine the cause of a plant problem when beginning with a nonspecific symptom is to find more symptoms. Gather more data. Produce a set of symptoms. Even if your set of symptoms includes nothing more than a group of individually nonspecific conditions, it might still lead to correct diagnosis of the cause or causes.

If you start with yellow leaves, for example, you could increase your symptom set by noticing:

1. Where on the plant the yellow leaves are.
2. When they first appeared.
3. What unusual cultural things might have happened.
4. What other leaf conditions you can see (browning? leaf drop? leaf size change?).
5. What soil conditions are present (wet? dry? compacted? acidic? salty?).

Perhaps the help of an outside diagnostic clinic or a soil testing service is necessary. The result of all this investigation will be an increased ability to correctly diagnose the cause of the symptoms and correctly plan a solution.

Diagnosing infectious diseases

Recognizing infectious diseases on flowers, leaves and stems is sometimes more difficult than recognizing insect or mite pests because you can't see the pathogens. You may be unfamiliar with the pathogens and the damage they cause. Most often, the pathogens you are looking for will be fungi.

A *fungus* often grows on a leaf similar to the way mold grows on bread or a rotted spot develops on a fruit or vegetable. Look for a circular spot. Sometimes one circular spot overlaps another, giving a blotchy appearance. Look for concentric rings in the spot, creating a bullseye. Sometimes you can see fungal spores on plant leaves or stems. Inspect for fluffy, moldy growth on the plant surface like those you see in powdery mildews. Also, look for black, pinpoint-like pustules within damaged tissue. These pustules are actually fungal formations in which many spores are produced and pushed to the outside. Many different fungal pathogens can occur on greenhouse-grown ornamentals, producing lesions in different sizes, shapes and colors.

Bacterial diseases sometimes appear as oily, greasy or water-soaked spots on leaves. The oilyness is often visible by turning the leaf over and viewing the lesion from the leaf's underside. Some bacterial diseases are systemic and cause

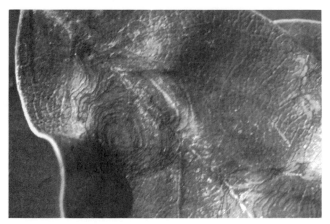

*Cylindrocladium on rhododendron. Note the circular spotting
with concentric rings.*

branch blighting, wilting and blackening. A common bacterial problem in
greenhouses causes a soft rot of plant stems that is sometimes smelly.

   Viral diseases often show up as yellowish or lighter green, ringed patterns
on plant foliage or as leaf distortions. The damage sometimes looks like damage caused by nutritional problems or herbicides.

   Infectious root rots can be diagnosed to some extent by directly observing
the root system. Off-color or brownish-to-blackish roots often indicate root rot.
Being able to pull off outer root tissue with your fingers (leaving behind the
stringlike cortex of the root) indicates root rot. To determine the health of a root
system, you should know what a healthy root system looks like. Healthy roots
are firm to touch and usually white.

   Diagnosing root health problems is difficult, largely because the visible
upper plant symptoms are reasonably nonspecific, even though they may result

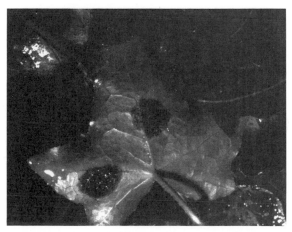

A greasy, water-soaked bacterial leaf spot on English ivy.

11

from poor root health. Plant wilting, yellowing, stunting, leaf scorch, leaf drop or death are symptoms of root health problems, but hundreds of problems with the root environment can be responsible. In these cases take your symptomatology further to get closer to the root environment for diagnosis. Soil tests, soil probing, drainage determinations, investigations of soil mixes, and microscopic observation or laboratory culturing of root tissue are all helpful methods to diagnose the problem.

Planning solutions to problems

Formulating the solution to a problem will be much easier if diagnosis is done carefully and correctly up to this point. Start by thinking generally about solutions and choose a plan from the array of plant cures and doctoring techniques used by the experts.

Four general approaches to the correction or management of a plant health problem exist. First of all, you can prevent the problems from showing up in the first place by proper health management. This is, of course, the best approach. Another general approach is to cure the sickness by correctly using a pesticide, by changing the environment or by pruning or removing the diseased plant part. Root systems damaged by poor drainage can sometimes be cured by repotting and reestablishing the roots in a more hospitable environment. This, however, is impractical with a large crop.

Generally, curatives are difficult because cures are generally applied to only one of the problem's many causes. For instance, spider mites may occur on a plant kept too hot and dry. Spraying the plant with a miticide will kill the mites, but will not cure the problem because you have not taken care of the underlying cause, the hot and dry weather conditions. The mites will soon return.

Another general approach to a plant health problem is to tolerate the situation. In many instances such tolerance remains the only alternative. In commercial greenhouses, however, tolerance isn't generally suitable.

Distorted growth and ringed spots caused by a virus on gloxinia.

Finally, you can simply start over: Remove plant material, discard it and bring new material. Again, this is a continuing and expensive general plant health management approach without some sort of determination and correction of the causes of the decline or health problem. On the other hand, in many situations starting over with new plant material is the only practical and economically feasible way to ensure plant health. For instance, you can best deal with bacterial wilt noted in only one of several geranium cultivars by completely discarding the suspect cultivar to protect the rest.

Remember, diagnosing problems on greenhouse ornamentals is a three-step process. First, you must correctly perceive the entire problem. Second, you must correctly determine the problem's causes, looking for environmental stresses as well as a pathogen or pest. Third, you must plan your course of action to solve the problem and prevent its continuing or happening again. Proper attention to each of these areas requires ordered thinking, data gathering and careful planning. Experience and knowledge are helpful, but good greenhouse production professionals know they can always increase their diagnostic skills. Thus, diagnosis is an important part of your continued success as an integrated plant health manager.

A general diagnostic guide for plants exhibiting nonspecific leaf symptoms

Symptoms	Possible causes
Brown or scorched leaf tips	Poor root health from overwatering, excessive soil dryness (especially between waterings), excessive fertilizer or other soluble salts in the soil
	Specific nutrient toxicities (such as fluoride, copper or boron)
	Excessive heat or light
	Pesticide or mechanical injury
	Air pollution
Leaf spots or blotches	Excessive soil dryness coupled with high temperatures
	Chemical spray injury
	Fungal or bacterial infections
Yellow green foliage:	
Older leaves	Insufficient fertilizer, usually nitrogen
	Poor root health due to poor drainage
	Insufficient light from shading
	Root rot diseases

Newer leaves	Soil pH imbalance Trace element imbalance
General yellowing	Too much light Insufficient fertilization High temperatures, especially when coupled with dryness Insect infestation or root rot disease
Yellowing foliage of one branch	Fungal or bacterial canker Injury Fungal infection of vascular system
Leaf drop	Poor root health from overwatering, excessive dryness, excessive fertilizer or other soluble salts in the soil Sudden change in light, temperature or relative humidity Insect or mite infestation
Wilting or drooping foliage	Poor root health from overwatering, excessive dryness, excessive fertilizer or other soluble salts in the soil, or poor soil drainage A toxic chemical poured into soil Root rot disease Fungal or bacterial cankers
Yellowed leaves with a tiny speckled spotting; leaves later bronzed and drying	Spider mite infestation Air pollution Insect infestation Low soil pH
Deformed or misshapen leaves	Herbicide injury Viral infection Insect or tarsonemid mite infestation

---

To more correctly determine causes of greenhouse plant problems, compile a symptom list. Look for specific symptoms associated with these nonspecific symptoms. In later chapters, specific symptoms will be noted for many pests and diseases on floral and foliage crops.

Chapter 3

Common Diseases, Insects and Mites

*T*he following list is by no means complete. Its intent is to help you become generally familiar with disease and pest problems you may encounter when growing flower or foliage crops in the greenhouse or outdoors and with some ideas concerning control. An understanding of these generalities may help you avoid plant problems and help you achieve integrated plant health management.

Common diseases and pathogens

Bacterial diseases

Bacteria comprise a diverse group of single-celled microbes that cause many plant diseases. Common diseases include bacterial wilt of carnation (*Pseudomonas caryophylii*); soft rot of cuttings, corms and bulbs (*Erwinia chrysan-*

Bacterial wilt on geraniums.

themi); bacterial leaf spots of foliage plants (*Xanthomonas* spp.); or crown gall (*Agrobacterium tumifaciens*).

Bacterial disease control is preventive, centered around cultural control measures. Always begin with clean plant material from a reliable source. Cultured cuttings are generally safest. Make sure cuttings are at optimum hardness for quick rooting. Keep plants from getting too cool during rooting.

If bacterial diseases have begun to cause trouble in the crop, be very careful to avoid spread. Don't splash water about during watering, space the plants and harden the crop by running it cool, dry and with reduced fertilization. Applying fixed copper sprays sometimes helps limit disease spread.

Nematode-caused diseases

Several types of tiny roundworms cause plant diseases on floral and foliage crops. Lesion nematodes (*Pratylenchus*) and pin nematodes (*Paratylenchus*) cause plant stunting and poor growth on many plants because their feeding weakens the root system. The root knot nematode (*Meliodogyne*) causes nodules to form on roots, impairing root function. This also causes stunting. The foliar or spring crimp nematode (*Aphelenchoides*) lives within the leaf tissues of many flower crops. It kills leaf tissue, resulting in brown lesions on older leaves.

Good sanitation is the primary control for these soil-borne pathogens. Soil sanitation kills both adults and eggs. Fumigants are as effective as steam for this purpose. After plants are growing, apply Oxamyl as granules or Vydate as a

Foliar nematode on a bird's nest fern.

drench. For foliar nematodes, Demeton sprays are effective (check whether the pesticide label allows this use).

Viral diseases

Viruses are disease-causing agents that live and multiply only within the host plant's living cells. They are most often spread by plant contact or by sucking insects such as aphids or thrips. The symptoms they cause are diverse, depending on the virus. Generally, vein banding, mosaic (a mixture of irregularly-shaped dark and light green areas on the leaf), flecking or ring spotting will show up on leaves. Sometimes growth abnormalities will appear similar to damage caused by weed killers. Finally, viruses can stunt plants.

Control viral diseases by using cultured cuttings when available. If you notice viral symptoms in a crop, try to prevent spread by controlling insects and avoiding unnecessary handling of the plants. Discard any plants showing symptoms.

Yellows diseases

Several flower and foliage crop diseases are caused by an unusual microbe called a mycoplasma. Mycoplasmas are very tiny with structure and functions between viruses and bacteria. Aster yellows is a typical, mycoplasma-caused plant disease. Another is the lethal yellowing disease of some palms. In these diseases, the plant develops growth abnormalities and a general yellowing. Flower tissue often aborts and reverts to a vegetative shoot condition. "Witches' brooms," a proliferation of buds and shoots coming from an older stem, often develop.

Mycoplasma diseases were once thought to be caused by viruses. Like some viruses, the causal agents are transmitted by leafhoppers or infected cuttings. Therefore, control involves controlling insect vectors and using only non-infected material for propagation. Also, weed control is important to eliminate reservoirs of insects and the pathogen. In palms, injecting older trees with

Ring spotting caused by a virus.

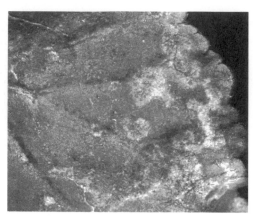

Typical powdery mildew lesions.

antibiotics has proven effective. Fortunately, mycoplasma diseases of floral and foliage crops are relatively rare, especially in greenhouses.

Powdery mildews

The powdery-mildew-causing fungi are very host specific, but widespread. Common hosts include roses, gerbera and chrysanthemums. The white growth appearing on leaves and stems is the fungus growing on the tissue surface. Small structures, called haustoria, grow within the host cells, injuring them as they obtain food. Powdery mildew usually doesn't kill a plant, but unsightly fungal lesions greatly reduce the plant or cut flower quality.

You can treat powdery mildew effectively with chemicals, eradicating it and "curing" the plants. The white lesions will remain, however, even though the fungus may be dead. Environmental control of powdery mildews can also be quite successful. In the greenhouse, reduce the high humidity that often occurs at night by venting and heating at sundown. Use fans to circulate the air but avoid drafts from open doors.

Rusts

Like powdery mildews, rusts are also host specific. Rusts usually sporulate abundantly on leaf tissue. The masses of orange to dark red spores show when plants become diseased. Rusts occur on many floral crops, notably geraniums, snapdragons, carnations, chrysanthemums and roses.

Rusts are, for the most part, cool weather diseases. Rust spores spread in air currents and in splashing water. They must have water to germinate and infect the leaf. Therefore, control involves keeping temperatures high, watering early in the day, spacing plants to allow more rapid drying and applying protective fungicide sprays.

Water mold root and crown rots

Pythium and Phytophthora are called water molds because they have a spore stage that spreads by swimming in water. These organisms attack a wide variety of plants, causing root rots, stem rots and cutting rots. Often they won't kill a

Rust lesion spores are common on lower leaf surfaces.

plant but will "prune" the root system, resulting in poor growth and yellowing or stunting of the plant top.

You can prevent water mold root rots with a good sanitation program. These organisms are probably found in all soils, so contamination can take place even though the planting media is sanitized. Often this contamination comes from unclean areas under greenhouse benches or near water lines. Improving the plants' growing media drainage environmentally controls these root rots. The fungi don't survive in well-drained media. Finally, you can apply soil drench fungicides.

Other root and stem rotting fungi (the imperfects)
Aside from water molds, many other fungi cause root and stem rots. The fungus Rhizoctonia lives in the soil and attacks a wide variety of crops. Some species

The plant on the right has root rot. Note the nonspecific, general stunting that has resulted.

Rhizoctonia growing on soil particles.

of Fusarium, Cylindrocladium, Sclerotinia and Thielaviopsis behave similarly. All of these fungi can persist in the soil for many months through specially adapted resting structures.

Control again involves soil sanitation programs. Do not plant cuttings too deep. Once the disease is present in a crop, apply soil drenches. The fungicides used will usually be different from those used for water molds.

Plant wilts

Fusarium oxysporum and Verticillium are two fungi that cause wilting on many floral crops. These fungi invade roots, grow into the stem and plug the plant's vascular system. Along with wilting, they often cause browning of the vascular system. They also have resting structures to help them persist from crop to crop. Generally, plant debris left after harvest contains these structures.

Chemicals can't control plant-wilting fungi well once they are in plants. Therefore, good soil sanitation before planting, along with using cultured cuttings when available are the best means of control. General stress management also helps.

Chrysanthemum vascular wilt caused by Fusarium.

Botrytis on geraniums.

Leaf and flower spots

Like rusts or mildews, these fungi are spread over long distances by airborne spores that land on plants and infect them. Most notable among this group is Botrytis. Septoria, Alternaria and Ascochyta also are commonly found on many foliage and floral crops. Leaf spotting fungi require water on leaf surfaces for infection to occur. They require splashing water for spore spread from leaf to leaf. Avoiding water on leaves and routinely spraying with protective fungicides are the best ways to control or prevent these diseases.

Cutting rot and damping off

The organisms responsible for cutting rots and damping off have already been mentioned. Cutting rots are often caused by water molds, bacteria, Rhizoctonia and Botrytis. Damping off of seed or seedlings often results from invading water molds or Rhizoctonia. Use the same controls you use for other diseases that involve soilborne organisms. In addition, keep cutting benches and germination flats warm. Place cuttings far enough apart to avoid rapid plant-to-plant spread. Propagation demands great attention to sanitation and other environ-

Fungal damping off of seedlings.

21

mental disease control methods. Don't use fungicides in propagation unless absolutely necessary.

Common insect and mite pests

Aphids (Homoptera)

Aphids are soft-bodied, sluggish insects that multiply rapidly. In greenhouses and other tropical areas, all aphids are females that produce live young (nymphs). Each female aphid can produce 50 or more nymphs during her lifespan, and under average greenhouse conditions, nymphs can mature and begin reproducing in about seven days.

Aphids are sucking insects that insert their mouthparts into plant tissue and extract plant fluids. These insects produce a sticky honeydew. They transmit plant viral diseases, and feeding by high aphid populations can weaken plants. Their nuisance value, however, usually exceeds their actual crop destruction potential. Although pesticide resistance is widespread in this group, and some predators and parasites are available, aphid control on ornamentals depends almost exclusively on insecticides. Early aphid detection is very important in managing them.

Whiteflies (Homoptera)

Whiteflies are common pests on many floral crops. Two main species, the greenhouse whitefly and sweet potato whitefly, cause most problems. As with aphids, whiteflies have piercing, sucking mouthparts. Adult and nymph feeding produces honeydew. All developmental stages (including eggs, nymphs, "pupae" and adults) usually occur on leaf undersides. Whiteflies complete their life cycle in 21 to 36 days, depending on temperatures.

A few whiteflies can be a nuisance, but heavy populations can weaken plants. The honeydew can be a substrate for a black, sooty mold. Besides making a plant unsightly, the mold interferes with the plant's photosynthetic processes and can reduce vigor or even cause death. The sweet potato whitefly is capable of transmitting many viral diseases.

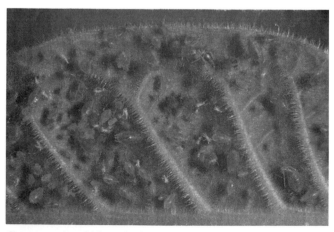

Green peach aphids on leaf underside.

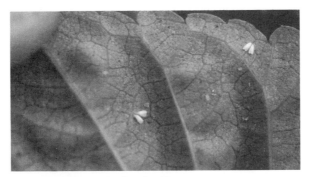

Adult whiteflies on leaf underside.

Sooty fungus growing on honeydew. Whiteflies, aphids, mealybugs and soft scales all produce honeydew.

Whiteflies have become tolerant of or resistant to several pesticides, but pesticide application still is the primary control method. Potential biological control on some crops includes using parasitic wasps, predators or fungi.

Scale insects and mealybugs (Homoptera)

Several species of "hard" and "soft" scale insects and mealybugs can occur on greenhouse plants. As with aphids and whiteflies, scale insects and mealybugs have piercing, sucking mouthparts. Only soft scales and mealybugs produce honeydew. True "scale" insects have a covering that protects them from certain pesticides and other environmental hazards. They remain immobile, except for the first-instar nymph stage ("crawler"). Mealybugs are mobile during all stages except the egg stage.

Infestations often occur when infested plants are brought into the greenhouse. Low numbers of these insects usually are quite inconspicuous, and by the time an infestation is obvious, control is very difficult. Growers should thoroughly inspect new plants before placing them in the main greenhouse range.

Soft scales on leaf underside.

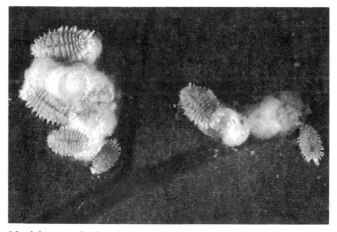

Mealybugs on leaf underside.

Thrips (Thysanoptera)

Thrips are small, slender insects, less than one-eighth-inch long, that feed on flowers, foliage, or pollen. Thrips feed by rasping plant tissues with a portion of their mouthparts and sucking plant sap with another portion. Leaf and flower deformity are injury symptoms. Black dots of excrement also are clues to thrips infestation. Several common thrips species, including the western flower thrips and onion thrips, transmit the virus causing spotted wilt, which can affect numerous ornamental crops.

Thrips are difficult to control because they infest flowers, often in the tight bud stage. Frequent pesticide applications are necessary to reduce populations, but pesticide resistance is widespread. No biological control is used on commercially produced ornamentals, although some experiments using predatory mites or predatory bugs have been quite successful.

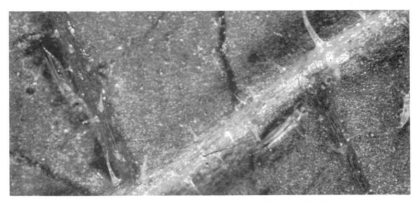

Adult (left and center) and immature (right) thrips on leaf underside.

Leafminers *(Liriomyza sp.) (Diptera)*

Leafminers are larvae of small, black and yellow flies. The larvae cause most injury by feeding between upper and lower leaf surfaces, leaving narrow, winding trails or mines. Adults can also cause noticeable injury, making small feeding and oviposition punctures in the leaf surface that soon turn white, giving leaves a speckled appearance.

Pesticide resistance is also a problem in leafminer control. Recent new product registrations have, however, made control with pesticides possible. Biological control with parasitic wasps is a real possibility on some crops.

Fungus gnats *(Diptera)*

Fungus gnats are small, fragile, dark gray or black flies often seen running over the soil surface, especially wet areas. Seedlings, rooted cuttings and young plants have been damaged by fungus gnat larvae feeding on root hairs or roots.

Heavy leafminer larval injury to chrysanthemums.

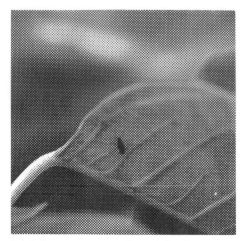

Adult fungus gnat on poinsettia leaf.

In addition to physical root injury by larvae, the presence of the midgelike adults is objectionable in retail shops, homes and hospitals. Plants produced in growing media containing a high percentage of organic matter have more problems with fungus gnats.

Most control programs for fungus gnats involve pesticide drench applications against larvae. Biological control includes using nematodes or predatory mites.

Caterpillars (*Lepidoptera*)

Many species of moth larvae are present on floral crops. Most moths are nightflying and are attracted into production areas by lights. They lay eggs singly or in masses, depending upon the species. Larvae have chewing mouthparts and can consume leaves and flowers. Some species bore into stems or tie leaves together with silken strands.

Fungus gnat larvae.

Caterpillar on geranium leaf.

Although the group shows pesticide resistance, insecticide control is certainly possible. Microbial insecticides often successfully manage certain species.

Spider mites *(Acari)*

The two spotted spider mite, *Tetranychus urticae*, attacks a wide range of greenhouse crops. These mites' pesticide resistance has increased their importance in recent years.

All stages remove plant sap, generally from lower leaf surfaces, giving the upper surfaces a characteristic spotted or lightly mottled appearance. During heavy infestations, webbing may cover the entire plant.

Spider mites do best at high temperatures and low humidities. During such times, population "explosions" can occur. Certain pesticides (for example, some pyrethroid insecticides) can stimulate spider mite reproduction. Biological control using predatory mites may be practical on certain crops.

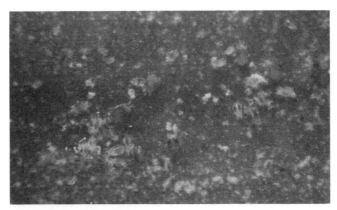

Two-spotted spider mites.
Photo courtesy of National Foliage Association.

A pest and disease diagnostic guide

You will recall in our chapter on diagnosis we stressed the importance of compiling enough information about a problem to be sure you're diagnosing it correctly. If you don't carefully work through a diagnosis, trouble can arise from many directions. Diagnostic carelessness may cause you to react improperly to a non-specific symptom. For instance, you might conclude that yellowed leaves are caused by lack of fertilizer and you fertilize the crop. If the cause were actually Pythium root rot, you would have made a costly mistake indeed, for the fertilizer would greatly worsen the root stress.

Another problem arising from incomplete diagnosis is not understanding specifically what pests or pathogens are present, how they got there, and what is favoring their development. Many pests are transported great distances on cuttings or small plants. Close inspection of newly arrived plant material is a basic part of your plant protection program. You need to know where to look and what to look for. For instance, the "typical" adult of various insect pests may not be present. Feeding scars or eggs may be all that you will find.

The following pages illustrate some views of pests and diseases that can help you increase your diagnostic skills. This is not a complete guide. Don't attempt to find all the pests and diseases of floral and foliage crops neatly illustrated for you to simply match to your material. That is not diagnostics. That is not the way to become good at diagnostics. Use this guide to stimulate your thinking about finding out what is wrong with plants when the time comes for you to have to do so.

Powdery mildew fungus on chrysanthemum stems. On some mum cultivars, stems are more severely affected than are leaves.

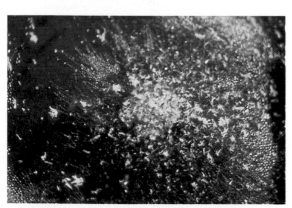

Powdery mildew lesion on a begonia. This magnified view was taken through a 30 X field microscope. Fungal growth strands radiate outward from the older part of the lesion. The accumulated material in the center is the spore mass.

Rust fungus on geranium leaf. The rust fungus bursts outward from within this leaf as the fungus produces more and more spores. You are seeing the spores. This happens more readily on the leaf underside and occurs on many plants.

Fungal infection of peperomia leaf. This brown, blighted area has concentric rings of discolored tissue and raised structures that bear the fungal spores. The spores are produced in tiny containers on the older portion of infected tissue.

Botrytis on petunia. Botrytis readily produces mold growth bearing brown spores. The growth can be seen over this petunia's infected crown tissue.

Botrytis blight on geranium. To confirm the diagnosis, look for the brown spores pictured in the photo. You may need a hand lens magnifier.

Botrytis blight on poinsettia. Botrytis blight of cuttings often results in a soft rot that becomes covered with a gray mold. The moldy growth engulfs all the plant tissue eventually, then spreads to neighboring poinsettia cuttings.

Botrytis blight on marigold. Botrytis often causes stem cankers, as on this marigold. Since there is no gray mold growth or brown spore production on the affected tissue, this case is difficult to diagnose properly. Take plants like this and put them in a plastic bag with a damp paper towel for two or three days. The fungus will start producing spores.

Dodder on impatiens. Dodder is a parasite that overgrows and strangles many ornamentals. Dodder produces flowers and seeds as it matures. Birds carry the seeds to new places.

Xanthomonas bacteria on begonia. The bacterium causing this blight infects the leaf through hydathodes at the leaf's edge. By noting the progression of symptoms, you can visualize how the pathogen is moving down the leaf veins toward the petiole. It continues into the stem, becoming systemic in the plant.

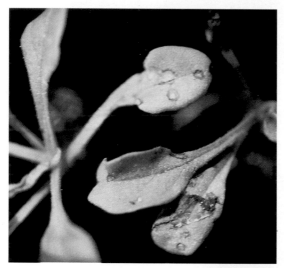

Pseudomonas bacteria on petunia. These bacterial spots are localized and not systemic. Note that they have a definite border and are scattered through the leaves.

Tobacco mosaic virus on impatiens. This systemic infection has stunted and malformed the plant. Any cuttings taken from the plant will be like this plant.

Tomato ringspot virus on geranium. The yellow spots are caused by tomato ringspot virus. The yellowing between the veins on some leaves isn't caused by the virus. Tomato ringspot symptoms of geranium normally go away as the weather warms and brightens.

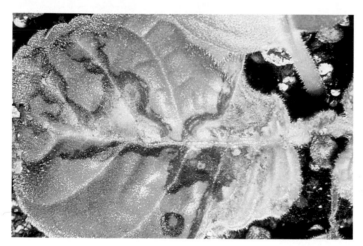

Tomato spotted wilt virus on gloxinia. Note how the growing point of the plant has become infected. This plant will die.

The black stem streaking on the exacum may be a result of tomato spotted wilt virus. Be careful, however. Other environmental problems can cause similar problems.

The fungi causing damping off of seedlings occasionally grow up onto the leaves and stems of fallen plants. This can help you determine if damping off is the problem. It cannot help you decide which fungus is involved, however, because many damping-off pathogens can do this. Send this sort of material into a plant clinic for a complete analysis.

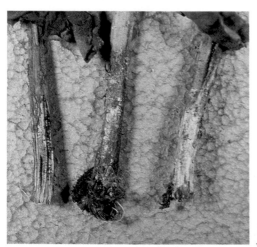

Rhizoctonia cutting rot on chrysanthemum. Rhizoctonia cutting rot typically results in a dry, stringy rot, such as on these chrysanthemum cuttings. If you look closely at the rotted tissue, you often see brown, fungal strands.

Rhizoctonia on poinsettia cutting. Rhizoctonia can rapidly attach a recently transplanted cutting, especially if it's placed too deeply in the pot. As poinsettias and other plants mature, they become resistant to Rhizoctonia infection.

Sclerotinia crown rot fungus on creeping phlox. In two or three days the white fungal tufts mature into brown structures that look like seeds. These structures can be transported many miles and can survive for years.

Phytophthora on African violet. Phytophthora root and crown rot attacks the underground plant parts rapidly and indiscriminately. The result on this African violet is a total plant collapse. The tissue you can see here can be lifted up, revealing the totally rotted tissue below.

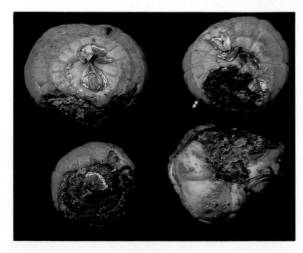

Fusarium storage rot on gladiolus corms. Note that this is a superficial rotting, not an internal, vascular infection. Many fungi can cause such superficial rotting. Laboratory examination is usually needed for complete diagnosis.

Localized injury of cyclamen. Symptoms of a problem on this cyclamen are evident on only part of the plant. You are correct to conclude that a localized injury or infection has occured.

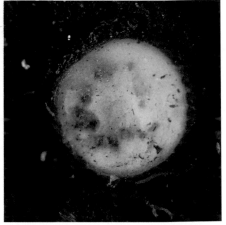

Fusarium on cyclamen. This is a cross cut through the corm from the plant in the above photo. Note how the veins in the corm in the area of the yellowed leaves are brown and diseased. Laboratory cultures of this tissue will indicate that Fusarium is present.

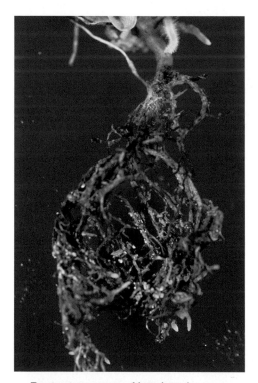

Root rot on pansy. Note how by washing the root system, definite blackened areas can be seen throughout the roots and even on the lower stem. These symptoms are typical of Thielaviopsis, the cause of black root rot disease.

An aphid colony on a flower bud. Aphids can be found on nearly all plant parts, but many species prefer terminal growth areas. Note the larger (adult) and smaller (nymph) individuals. Mature aphids are 1 to 2.5 millimeters long and variable in color.

Greenhouse whitefly adult on leaf underside. Adult whiteflies are usually found on upper leaf undersides. Because of plant growth, the immature stages are often seen on leaves well below the growing point.

Enlarged view of a female aphid and newly-produced nymph. In greenhouses and tropical areas aphids are generally all females and produce live young. Each female can produce about 50 offspring.

Sweet potato whitefly adults on leaf underside. Sweet potato whiteflies are usually smaller, more yellow-bodied, fly more readily and hold their wings more vertically when at rest than greenhouse whiteflies.

Greenhouse whitefly adults (two pairs) and a single sweet potato whitefly adult on salvia leaf underside. The size difference and wing angle are visible.

Encarsia formosa parasite on sweet potato whitefly nymph. E. formosa deposits an egg in whitefly nymphs, and the resulting parasite larvae develops inside the nymphs, killing them. Parasitized greenhouse whiteflies turn black and sweet potato whiteflies turn brown.

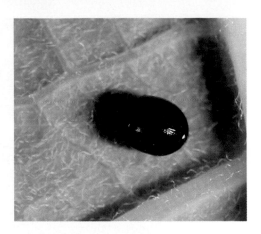

Delphastus pusillus, *a ladybird beetle predator of whiteflies. The adults and larvae are voracious feeders, consuming several hundred whitefly eggs and/or nymphs daily. Although it may be difficult to determine the head from the tail, the beetle's antennae are visible in its shadow at left.*

Adult female mealybugs (length about 3 millimeters), egg sacs and mealybug nymphs. Mealybugs can be found on all plant areas. Large populations often develop in leaf whorls and axils. Mealybugs produce large quantities of honeydew.

Soft scales on cyclamen leaf underside. Soft scales are about 1 to 4 millimeters long, and can be found on all plant areas. Soft scales, similar to mealybugs, whiteflies and aphids, produce large quantities of honeydew.

An armored scale (oystershell scale) on a plant stem.
Armored scales don't produce honeydew and are variable
in size and shape. Armored scales of this shape are often
3 to 4 millimeters long. The color is variable, depending on
the species.

Adult thrips on a rose bud. Thrips often
deposit eggs in and around developing
flowers, so when flowers open, large
numbers are already present. Feeding
damage will discolor and distort flowers.

Circular armored scales on a palm
leaf. Circular scales are 1 to 3
millimeters in diameter and variable in
color. Scale insects can develop very
high populations over time, mainly on
woody foliage plants.

An enlarged view of an adult thrips (1 to 2 millimeters long). Note the long, narrow, fringed wings.

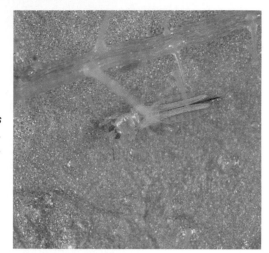

Thrips feeding injury on a chrysanthemum leaf. Although many species prefer flowers, they can survive quite well on foliage.

A thrips predator, Amblyseius sp., holding a thrips larva. Many mites in the family Phytoseiidae will feed on young thrips, but results of using predatory mites to control thrips in experiments and in commercial greenhouses have been mixed.

An Egyptian cotton leaf worm, Spodoptera littoralis, *on chrysanthemum. These insects occur mainly in the Middle East, but in this case they were shipped into the United States on plant cuttings. The size and injury are about the same as with beet armyworms.*

A beet armyworm, Spodoptera exigua *larva on chrysanthemum. These larvae, about 3 centimeters long when fully developed, are among the most serious pests of chrysanthemums and other crops. They eat leaves and flowers. Pesticide resistance is widespread.*

An adult leafminer, Liriomyza trifolii. *These small (3 millimeter) yellow and black flies can infest many ornamental and vegetable crops. Several species can cause problems, and have been sent through the world on infested plant material.*

An adult leafminer and feeding/egg-laying punctures in a chrysanthemum leaf. The leafminer females make holes in a leaf with the ovipositor. Eggs are inserted into some of the holes, while both sexes use other holes to feed on exuding plant fluids.

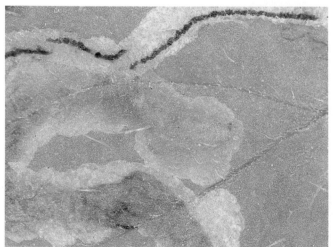

A fully developed leafminer larva at the end of a leaf mine. Larvae are about 8 millimeters long when fully developed. The black mouth hooks used by larva to produce the leaf mine are visible. The larva will exit the leaf mine and drop off the plant to form the pupal stage.

An adult fungus gnat on a pot. These dark adults, about 3 to 4 millimeters long, are often seen running over the growing medium surface, where they deposit their eggs. Adults do no direct plant injury but may carry root rot pathogens from infected to healthy plants.

A fungus gnat larva (center) on a poinsettia cutting. Fungus gnat larvae can be recognized by their shiny black heads. The larvae are about 8 millimeters long when fully developed. Plant growth can be reduced by the larvae feeding on plant roots.

An adult two-spotted spider mite female and egg on a leaf underside. Adult females, which predominate in mite populations, are about 0.5 millimeter long. High mite populations can kill plants by direct feeding injury. Lower numbers can disfigure plants by producing webbing.

Two-spotted spider mites and webbing on rose flower. The tiny mites and webbing are easily visible. Spider mites often gather at plant tops, particularly as plants mature.

Chapter 4

Plant Diseases and Control

*H*aving some knowledge of general plant pathology and its relationship to their agricultural specialty is important to growers. All plant species, both wild and cultivated, are subject to disease attacks. Over 80,000 different diseases have been recorded in the world; more than 50,000 occur in the United States. Each species or plant is subject to only a small number of these diseases. The kinds of diseases a particular plant might get depends on the plant type and the cultural or environmental conditions present. Refer to Appendix III for a list of commonly cultivated floral and foliage crops and potential diseases.

Cultivated plants are usually *more* susceptible to disease than their wild relatives—partly because large numbers of the same plant are often grown closely together. Disease-causing organisms (pathogens) often get established under these conditions. Once this happens they may spread rapidly. In addition, many of our valuable crop and ornamental plants are basically very susceptible to disease and would have difficulty surviving in undisturbed nature. Finally, cultivation—be it roses, cabbages or sycamores—is disturbing nature and tends to create environmental stresses. Stresses often weaken plants and subject them to pathogenic infections, as noted in Chapter 1.

Plant diseases aren't new. Fossils have proven that plants had disease enemies 250 million years or more before man appeared on earth. Plant diseases undoubtedly developed as life developed on earth. Many injurious pathogens including rusts, smuts, mildews, blights and blast are mentioned in the *Bible* and other ancient writings. These diseases and others have plagued man, causing famine and unrest since the dawn of recorded history. Plant diseases contributed to the defeats of Alexander the Great and the fall of the Roman Empire. Potato late blight caused the great Irish potato famine and emigration (1845 to 1860). The American chestnut tree has, for all practical purposes, been eliminated from our country because of a disease called chestnut blight. The southern corn leaf blight epidemic of 1970 cost U.S. farmers over 700 million bushels of corn.

Elatior begonia hybrid with powdery mildew.

What is plant disease?

When a plant is continuously disturbed over a fairly long period of time by some factor or group of factors in its environment that interferes with its normal structure, growth or functional activities, it is said to be diseased. Diseases can be infectious or noninfectious depending on the living or nonliving status of the causal factors. There is often no sharp distinction between healthy and diseased plants. Disease may be merely an extreme case of poor growth. Knowing normal growth habits, variety (cultivar) characteristics and normal variability of plants within a species helps you recognize a diseased condition.

Diseases are commonly named on the basis of the most obvious symptom. Some diseases result in plugging of the water-conducting vessels in a plant, producing a wilting condition similar to drought. Root rots destroy the feeding roots that absorb water and nutrients from the soil. Leaf spotting or blighting diseases reduce photosynthesis, resulting in less food manufactured by the plant. Seeds, corms, bulbs, seed pieces, fruits and flowers may be destroyed by rots, scabs or blights. These diseases reduce a species' reproductive ability.

What causes plant diseases?

Some plant diseases result from unfavorable growing conditions (called disorders, or nonparasitic, noninfectious, abiotic or physiogenic diseases). Others result from living pathogens (called infectious or parasitic diseases). Living pathogens include bacteria, fungi, viruses, viroids, mycoplasmas, nematodes and parasitic seed plants. You must know the disease cause because control measures vary widely from one to another.

Unfavorable growing conditions

Noninfectious diseases can result from temperature extremes, an excess or deficiency of light, water or essential soil elements (for example, nitrogen, phosphorus, potassium, calcium, iron, magnesium, manganese, boron, copper, molybdenum, zinc, sulfur), unfavorable soil moisture and/or oxygen relationships, extreme soil acidity or alkalinity, pesticide or fertilizer injury, air or soil toxic impurities, mechanical and electrical agents, fire, soil compaction plus unfavorable postharvest or storage conditions for fruits, vegetables or nursery stock. Poor plant health (disease) is more often a result of unfavorable growing conditions than of disease-producing organisms. Noninfectious troubles *do not spread* from sick to healthy plants. They often arise, sometimes very suddenly at about the same time, on several plants growing in an area or environment.

Pathogens

Infectious diseases are generally caused by parasitic microorganisms (bacteria, fungi, nematodes, mycoplasmas, viruses and viroids). These organisms or agents attack plants and live in or on the plant and at its expense. The infected plant is called the host plant. These pathogenic parasites cause infectious diseases that often spread easily from diseased to healthy plants.

Bacteria. Most of the 4,000 bacteria species that have been described are either harmless or beneficial to man. However, over 100 kinds are known to cause human and animal diseases, and more than 200 kinds incite diseases in plants. Bacteria are one-celled microorganisms found in all types of air, soil and water and are common on and in all plants, animals and man. They are visible only with a good microscope. Placed end to end like bricks, about 10 to 20 thousand bacteria measure 1 inch.

Bacteria enter plants through wounds (produced by adverse weather, man, insects or nematodes) and small natural openings such as stomates, lenticels, hydathodes, nectaries and leaf scars. Bacteria may multiply rapidly inside a plant, where they can kill cells (necrosis) or cause leaf spots or abnormal growth (tumors), block water-conducting tissues (wilting), or break down tissue structure (soft rot). The bacteria often migrate throughout the plant. Some produce toxins that poison the plant and produce yellowing (chlorosis), water-soaking and other symptoms.

People spread bacteria through cultivating, harvesting, grafting, pruning and transporting diseased plant material such as seed, tubers, corms, bulbs, cuttings and transplants. Animals (including insects, mites and nematodes), splashing rain, flowing water and wind-blown dust are other common disseminating agents.

The most common plant bacterial diseases are soft rots, leaf spots or blotches, leaf and stem blights, stem cankers, wilts and galls. Most bacterial diseases favor warm, rainy weather. They are common on overhead-irrigated greenhouse crops, especially in the South.

When conditions are unfavorable for growth and multiplication, bacteria lie dormant on or inside plants, seeds, storage organs (for example, flower bulbs, corms, rhizomes), plant refuse, garden tools and farm implements, or in soil. A few live several months or longer in bodies of living insects. An example is Stewart's disease on corn where the causal bacteria overwinter in the corn flea beetle.

Most bacteria that cause plant diseases are quickly killed by high temperatures (10 minutes at 125 F), dry conditions and sunlight. Many pathogenic bacteria in soil are eaten by minute animal life in the soil or are inhibited by

Bacteria masses oozing from a wilted geranium's cut vascular tissue.

antibiotic substances liberated by other soil inhabiting organisms—chiefly bacteria, actinomycetes and fungi.

Fungi. Like bacteria, fungi are simple, usually microscopic plants that lack chlorophyll (green pigments), so they can't make their own food. Over 100,000 fungal species have been described, probably less than half the total number in the world. Approximately 20,000 different fungal species cause the majority of infectious plant diseases, including all rusts, smuts and mildews; most leaf spots, cankers, scabs and blights; root, stem and fruit rots, wilts; and galls.

Fungi that grow on or in a living plant and obtain nourishment from it are called parasites, and the living plant is called the host. Fungi that feed on manure, dead leaves, stems, wood and other nonliving organic matter are called saprophytes. Many fungi parasitically attack living plants at certain times, yet live as saprophytes in plant debris or in soil at other times. Some fungi are obligate parasites, growing and reproducing only on living plants.

A typical fungus begins life as a microscopic spore with a function comparable to that of a seed in a higher plant. Under moist conditions the spore may germinate and produce one or more branched, tubular threads called hyphae. The hyphae grow and branch to form a fungal body called a mycelium. The mycelium may be an interlacing tangle of hyphae, a loose woolly mass, or even a compact, solid body. A parasitic mycelium may grow on the plant host's surface, appearing as delicate, whitish, cobweblike threads (powdery mildew) or as sooty-brown to black filaments (sooty or black mold). The mycelium may also be completely within the host plant (wilt-producing or wood-rotting fungi) and not evident on the plant surface.

Fungal hyphae may penetrate a plant in three ways:

1. Growing into a wound made by a tool or farm implement, hail, wind, blowing sand, insects, nematodes or other fungi;
2. Through a natural opening; or
3. By forcing their way directly through a plant's epidermis by combining pressure and enzyme action.

A powdery mildew spore about to be released into the air.
Photo courtesy of American Phytopathological Society, *Compendium of Rose Diseases.*

Spores are important to fungal multiplication, dissemination and survival. Spores are easily carried by air currents, splashing or flowing water, insects, mites, birds, slugs, spiders and other animals, plant parts (for example, seeds, bulbs, transplants, cuttings) and cultivating, harvesting and pruning equipment. Man also spreads spores on hands, clothing and shoes.

Certain fungal spores can blow a thousand miles or more, sometimes at altitudes up to 90,000 feet, before descending (frequently in a rainstorm) and infecting plants. Wet, sticky, hairy or rough foliage "traps" more spores than dry, smooth plant surfaces.

Some fungal spores are called resting or resistant spores. Resting spores allow certain fungi to withstand unfavorable growing conditions such as extreme heat, cold, drying and flooding. Spores of certain fungi may lie dormant in soil for many years, making them extremely difficult to kill.

Some fungi don't produce spores. They either multiply and overseason by forming compact masses of hyphae called sclerotia, or the fungal body divides into fragments that are broken off and spread by water, wind, man and other agents.

Fungal diseases, like bacterial diseases, are more prevalent in damp areas or seasons than in dry ones. Fungi that infect leaves, stems, flowers or fruit generally require a wet host surface during spore germination and penetration. Moisture is also essential to their rapid reproduction and spread. Soil-inhabiting fungi may or may not require wet soil.

Fungi, like bacteria, survive apart from host plants on and in plant refuse, soil, perennial plants (weeds), seed, tubers, bulbs, corms or occasionally in insects. Knowledge of fungal habits guides the development of effective control measures. For example, some root-rotting fungi are soil invaders but can't survive in soil for long periods without a host plant. Clean cultivation and crop rotation can often control these soil invaders, but can't control soil inhabitants. They can live for a long time in soil without a host plant. Sanitizing or chemi-

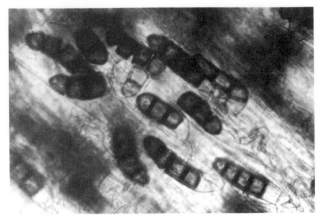

Thick-walled, resistant spores of the black root rot pathogen in a pansy root.

cally protecting seeds, bulbs, tubers, corms, transplants and plant roots is often needed to manage these pathogens and resulting diseases.

Viruses. The 300 or more plant-infecting viruses are complex, macromolecules that infect, replicate, mutate and otherwise act like living organisms only when in living plant cells. They are obligate parasites. With few exceptions, viruses are composed only of simple genetic material with a protective, protein "overcoat." Much smaller than bacteria (perhaps two-hundred-thousandth of an inch long), they can be seen only with an electron microscope. Viruses can truly be thought of as infectious genes.

Viruses cause plant disease by imposing a different set of genetic information on the host plant cell's biosynthetic apparatus. This results in altered and detrimental host metabolism causing invaded plant cells to reproduce more viral particles.

Symptoms of a given virus may differ between varieties or cultivars of the same species or even in the same plant (for example, tomato spotted wilt virus). Moreover, two unrelated viruses may cause identical symptoms in a given plant variety.

The most common virus-caused diseases classified by symptoms are mosaics, yellows, spotted wilts, ringspots, stunts, mottles and streaks. Depending on environmental conditions, many crop plants and weeds may harbor viruses but show no external symptoms, particularly at temperatures above 85 F. Viral diseases are often confused with nutrient deficiencies and imbalances, pesticide injury, insect-induced toxemias, mite feeding and even some genetic mutations.

A few plant viruses, such as tobacco mosaic virus (TMV), are quite infectious and spread easily from diseased to healthy plants by mere contact (for example, on worker's hands or clothing). Others are transmitted in nature only by insects (primarily by over 100 species of leafhoppers and 200 species of aphids, a few by thrips, mealybugs, whiteflies, planthoppers, grasshoppers, scales and certain beetles). Practically all viruses can be spread by vegetative

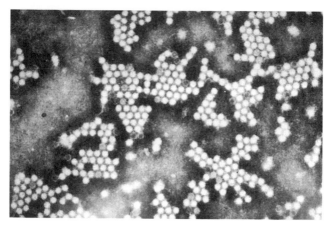

Viral particles greatly magnified by an electron microscope.

propagation (for example, grafting or budding, cuttings, seed pieces, root and rhizome divisions, tubers, corms, bulbs) or by parasitic plants (for example, dodder). Some 50 to 60 viruses are disseminated by infected seed and a few by mites, pollen, slugs and snails, fungi, nematodes or possibly other minute animal life in the soil.

Viral diseases are generally most serious in crops vegetatively propagated, including many ornamental flowers, bulbs or foliage crops.

Viruses often overwinter in biennial and perennial crops and weeds, insect bodies and plant debris. Once infected with a virus, plants normally remain infected for life. Most, if not all, plant-infecting viruses can "live" in a number of different plant hosts. Many viruses may be symptomless in most hosts.

Viroids. Several diseases (including chrysanthemum stunt) previously thought to be caused by a virus, have been found to be caused by a new class of infectious particles named viroids. The viroid is 80 times smaller than the smallest known virus. It consists of tiny genetic material with no protective protein coat. Like viruses, viroids invade plant cells and disrupt their functions.

Mycoplasmas. Over 60 of what used to be considered "yellows" or "witches' broom-"type viral diseases are now known to be caused by peculiar, one-celled, free-living organisms known as mycoplasmas. These are the simplest and smallest known organisms that can be grown in laboratory media free of living tissue. Mycoplasmas are saclike, without rigid cell walls. Smaller than bacteria but within the size range of large viral particles, mycoplasmas reproduce by budding (like yeasts) or binary fission (like bacteria).

Plant-infecting mycoplasmas have also been found in leafhoppers, the principal or only transmitters of certain "yellows" diseases. Future discoveries will probably show more leafhopper-transmitted diseases to be caused by mycoplasmas.

Nematodes. Nematodes are probably the most numerous multicellular animals on earth. More than 9,000 species have been described; a pint of soil may

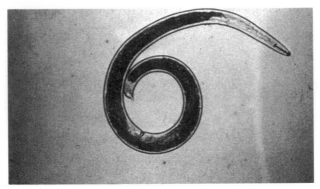

*A plant parasitic nematode showing the stylet in the head
(to the right).*
Photo by Richard M. Riedel, Ohio State University.

contain 20,000 or more nematodes. Most types are harmless, feeding primarily on decomposing organic material and other soil organisms. Several hundred are even beneficial to man since they are parasitic on soil pests. Plant-parasitic nematodes aren't easily visible to the naked eye because they're transparent and small (one-tenth to one-hundredth of an inch long).

Plant-parasitic nematodes obtain food by sucking juices from living plants. They feed through a hollow, needlelike mouth called a stylet. When feeding, the nematode pushes the stylet into plant cells and injects an enzyme mixture that predigests plant cell contents. The liquefied contents are then drawn back into the nematode's digestive tract through the stylet. Nematode feeding lowers a plant's natural resistance, reduces its vigor and yield, and affords easy entrance for wilt-or root rot-producing fungi, bacteria and other nematodes. Nematode-damaged plants are frequently more susceptible to winter injury, drought, disease and insect attack.

Plant-parasitic nematodes live freely in soil around roots or in fallow gardens and fields. In causing disease, the nematodes tunnel inside plant tissue (endoparasites) or feed externally on the root surface (ectoparasites). They enter plants through wounds, natural openings or by penetrating between root cells.

All plant-parasitic nematodes (about 1,800 species) require living plant tissues for reproduction. They lay eggs in or on plant tissues, especially roots, or in soil. Eggs hatch, sometimes after months or even years, releasing young, wormlike larvae (more accurately called juveniles) that usually start feeding at birth.

Most plant-parasitic species require 20 to 60 days to complete a generation from egg through four larval stages to adult and back to egg again. Some nematodes have only one generation each year but still can produce several hundred or more offspring.

Many environmental factors affect plant-parasitic nematodes' soil populations and development. Certain nematodes live strictly in light, sandy soils.

Some build up high populations in muck soils. High nematode populations and the resulting high crop damage are much more common in light sandy soils than in heavy clay soils.

Many plant-infesting nematodes become inactive at 40 F to 60 F and over 85 F to 105 F. The optimum temperature for most nematodes is 60 F to 85 F, but that varies greatly according to the species, developmental stage, activity and plant host growth.

Many nematode species are killed easily by air-drying soil after harvest or before planting. Some species remain alive but dormant under the same conditions. When dormant (in cyst or egg stage, or when embedded in plant tissues), they are much more difficult to kill with chemicals or heat than when moist and actively moving.

Any force or object that moves infested soil, plant parts or contaminated objects can easily spread nematodes. These include tools and machinery, containers, flowing water, wind, clothing, shoes, land animals, birds and infested nursery stock (especially plants with soil-covered roots).

Most plant-parasitic nematodes cause general poor root health symptoms. Root-knot nematode, the best known nematode, produces galls or knots on over 2,000 kinds of plant roots. Most root-feeding species, however, cause no specific symptoms. Infested plants are weakened and often appear to be suffering from drought, excessive soil moisture, sunburn or frost, mineral deficiency or imbalance, insect or mite injury to leaves, roots or stems, or diseases (for example, wilt, dieback, crown rot or root rot). Other common symptoms of nematode injury include stunting, green color loss and yellowing, twig and shoot dieback, slow general decline, temporary wilting on hot, bright days and lack of normal response to water and fertilizer. Reduced feeder root systems may be "stubby" or excessively branched and often shallow, discolored and decayed. Many times the first nematode injury sign in a field or greenhouse is scattered, circular-to-oval or irregular areas of stunted plants with yellow or bronzed foliage.

Growers can reduce losses caused by nematodes by watering during droughts, fertilizing to promote vigorous growth, practicing clean cultivation (good weed control), fall and summer fallowing, planting resistant varieties, plowing out susceptible plant roots right after harvest to expose them and nematodes to the drying of sun and wind, using heavy organic mulches, planting cover crops and starting with nematode-free planting materials. Crop rotation is often an important control measure. Because complete control is impossible or unlikely, periodic checks of nematode populations are necessary to prevent recurrent damage.

Parasitic seed plants. Many flowering, seed-producing plants are important parasites on ornamental plants. Among the more important are American (true) and dwarf mistletoes, broomrapes, witchweed and dodder. Like most higher plants, they reproduce by seed and spread from place to place by land animals, birds, wind, water, soil or as contaminants in seed lots.

Dodder is widely distributed. Its 170 species are orange-to-yellow, "leafless," threadlike vines that occur in tangled, yellowish-orange patches in fields

Dodder, a parasitic seed plant, on impatiens.

and gardens. The vines twine around field crops, vegetable and ornamental host plants, drawing them together and downward. A dodder-infested area is usually less than 9 or 10 feet in diameter the first year, but spreads more rapidly in succeeding years. Dodder seeds, rough, irregularly round and flat-sided and gray to reddish brown, are widely distributed as a contaminant in clover, alfalfa, sugarbeet and flax seed. To control dodder, plant certified, properly cleaned seed. Destroy infested areas well before the dodder plants flower. Upon drying, sprinkle infested patches with fuel oil and burn them. Spot treatment with a selective herbicide or soil fumigant is sometimes effective. Planting known, dodder-infested areas with resistant plants is another control method.

Conditions necessary for plant disease

Infectious diseases vary greatly in prevalence and severity from year to year and from one area to another. At least three conditions are necessary for disease development:

1. The air and soil environment (principally the amount and frequency of rains or heavy dews, relative humidity, air and soil temperatures and plant nutrients) must be favorable for infection;
2. The host plant must be susceptible; and
3. A virulent, disease-producing agent or pathogen must be present.

All three basic ingredients, commonly called the "disease triangle," must preceed infectious disease development.

Environmental factors and infectious disease development

Some important environmental factors that commonly affect the development, prevalence and severity of plant disease include temperature, relative humidity, soil moisture, soil reaction (pH), soil type and soil fertility.

Temperature. Each pathogen has an optimum for growth. Each fungal growth stage (for example, spore production and germination, growth of mycelium) may have a different optimum temperature. Knowing these temperatures, usually combined with suitable moisture conditions, permits forecasting devel-

INFECTIOUS DISEASES

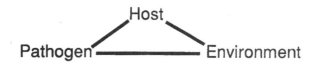

The disease triangle.

opment of diseases such as potato and tomato late blight, lima bean downy mildew, pome fruit fire blight and sycamore anthracnose. After planting, low soil temperatures after planting favor most seed rots and seedling blights. High temperatures (above about 85 F) commonly mask certain viral and mycoplasma disease symptoms, making them difficult or impossible to detect. Warm, dry weather favors insect build up, including aphids, leafhoppers and thrips. Virus diseases spread by these insects are usually more severe during warm, dry seasons.

Relative humidity. High humidity (88 to 100 percent) favors development of most leaf, stem, fruit and flower diseases caused by fungi and bacteria. Moisture from rain, irrigation or dew is generally needed for fungal spore germination, bacteria multiplication and penetration, and initiation of the infection process.

Soil moisture. High or low soil moisture may limit certain root rot diseases. "Water mold" root rotting fungi, such as Aphanomyces, Phytophthora and Pythium, thrive where soil is saturated or nearly so. Excessive watering lowers the soil's oxygen level and raises its carbon dioxide content, making the roots more susceptible to root rotting organisms. Diseases most severe under low soil moisture levels include Rhizoctonia damping-off of bedding plants, common scab of potato and white rot of onion.

Soil reaction (pH). Soil acidity or alkalinity greatly influences a few diseases such as common scab of potato and black root rot of pansy or poinsettia. The potato scab organism is suppressed by a pH of 5 to 5.2 or lower. Some growers add sulfur or acid fertilizer to keep the pH around 5. Likewise, black root rot can be almost totally suppressed by keeping growing media pH below 6.5.

Soil type. Soils high in loam or clay, or soils that are light (warm up quickly) favor certain pathogens. For example, Fusarium wilt diseases and nematodes are most damaging in lighter and coarser soils that warm up quickly in the spring. Fusarium wilt of potted chrysanthemums can be severe in the summer, when heat adversely affects the temperature of a light growing medium.

Cleaning and disinfesting benches between crops are part of good sanitation practices.

Soil fertility. Raising or lowering certain essential nutrient element levels influences development of some infectious diseases. Excessive nitrogen fertilizer increases destructiveness of fire blight on pome fruits and related ornamentals, Botrytis blights, bacterial soft rots of floral crops, Septoria and many other leaf spotting fungi, Sclerotinia dollar spot, snow molds and red thread of turfgrasses. For some bacterial diseases of foliage plants, it's been shown that increasing nitrogen fertilization can decrease the host's susceptibility.

Soils maintained at highly productive levels by the proper fertilizer use (based on a soil test), incorporating organic matter and using mulches tend to produce vigorous plants. These soils will be more resistant to many infectious diseases.

Plant disease control

Effective disease control measures are aimed at "breaking" the environment-pathogen-host plant triangle. For example, breeding a resistant variety can make a plant less susceptible. Changing the environment to favor host plant growth can hamper development, reproduction and spread of the disease-causing agent or pathogen. Incorporating these basic control methods into numerous practices helps check diseases.

Successful disease control is based on accurate diagnosis of the cause, thorough knowledge of the pathogen and its disease cycle and of how the host and pathogen interact with various environmental factors. Control should start with the purchase of the best varieties, seed and planting stock available. Disease-controlling practices must continue in the seedbed or nursery, throughout the field or garden season, and after harvest until the crop is consumed or fully utilized.

The most important point in controlling a plant disease is *choosing the best method* for a given situation. The best control for one disease on a certain host may not be the best method for another disease on the same or a different plant. Also, several control measures (integrated control) are often needed. For example, several cultural control practices are often combined with a protective fungicide spray or dust program. The best disease control program is *disease prevention*.

Disease control practices should be integrated into a broad program of biological or cultural and chemical methods needed to control the various pests—insects, mites, rodents, weeds—that attack a given crop. The next chapter discusses integrated disease control procedures for floral and foliage crop growing.

As you read the next chapter, remember that infectious diseases are controlled by one of four basic methods:

1. *Exclusion*. Disease-causing organisms and agents can be excluded from certain areas or countries. Quarantines and embargoes, inspecting and certifying seed and other plant materials, disinfecting plants, seeds and other propagative parts by using heat or chemicals are all methods to prevent pathogen movement into new areas.

   Selecting planting sites unsuited to disease development can help exclude plant diseases. Other exclusion practices include controlling storage and moisture conditions for vegetables, fruits and nursery stock, crop rotation with unrelated plants, sowing early or late and at the proper depth, and propagating and planting only disease-free material.

   Sanitation in and around propagating beds, greenhouses, gardens and fields is an excellent disease-exclusion measure. It involves cleaning and disinfesting potting benches, soil bins, head houses, greenhouse benches, tools and equipment. Disinfesting tools and equipment is most necessary if you have used them with diseased plants or around general debris.

2. *Protection*. Protect plants by uniform and timely applications of recommended, disease-control chemicals (fungicides, bactericides, nematicides); by following suggested cultural practices (for example, proper spacing, correct timing of planting and harvest, careful handling during harvest, grading and packing, proper pruning, watering and fertilization); and by altering the air and soil environment to make it unfavorable for the pathogen to infect, develop, reproduce or spread. Apply fungicides on the plant surface or in the soil where and *before* infection takes place. Most fungicides are protectants, not eradicants.

3. *Resistance*. The ideal disease-control measure is growing resistant or immune varieties, cultivars or species. Unfortunately, no plants resist all diseases. Developing resistant or immune varieties is a continuing process and critically important for low-value crops where other controls are unavailable or impractically expensive. Developing resistant varieties and cultivars is greatly complicated by pathogens having several to hundreds of physiological races. Most "resistant" cultivars or varieties ward off only one pathogen or pathogen race.

4. *Eradication*. Plant pathogens can be eliminated (eradicated) and in-sect transmitters can be controlled by pesticides—such as soil fumi-gants—by heat treatment, or by removal and destruction of diseased plants (roguing) or plant parts (surgery). These plants may be weeds or alternate host plants. Another eradication method is crop rotation which "starves" out soil-invading pathogens. Eradication by itself is rarely successful in controlling floral and foliage crop diseases.

Chapter 5

Integrated Infectious Disease Control

Managing leaf, stem and flower pathogens

As you have seen from the preceding chapter, many common pathogens attack plants, stems, leaves and flowers. These pathogens include those that cause powdery mildew diseases, Botrytis diseases, rust diseases, alternaria leaf spots, septoria leaf spots, bacterial leaf spots and wilts, viral diseases and Fusarium stem rots. The number of different organisms and the plants involved is truly large when we speak of these plant diseases.

Management strategies for stem, leaf and flower disease can be confusing if you think constantly about the many fungi, bacteria and host plants involved. Look for common denominators concerning the diseases. As you read in the preceding chapter, begin by thinking about the basic definition of a plant disease. A plant disease requires a host, a pathogen and an environment that allows pathogen interaction on the host. All three elements must be present at the same time, in the same place. Disease management strategies are designed to prevent this triangle of necessary elements. Finally, don't forget that, since plant health is an integrated and holistic concept, plant health management must also be integrated and holistic.

Sanitation

Sanitation is partly effective in controlling pathogens attacking stems, leaves and flowers. These pathogens often have airborne spores or inoculum that are airborne. Thus, they come in from fields or yards near your fields or greenhouse. You cannot sanitize your entire neighborhood. On the other hand, many pathogens depend on splashing water to spread. Destroying nearby plant debris will help control them. For instance, Botrytis is a fungus that will proliferate and produce spores on crop debris. Ridding your area of diseased plants is effective disease control to some degree. Never dump plant material under the greenhouse bench or at the side of a field and allow it to sit there and rot.

Botrytis producing spores on old, dead geranium leaves.

Many stem, leaf and flower diseases get into a crop through infected stock. This is particularly true for bacterial diseases, Fusarium wilt diseases, viral diseases and rust diseases. These pathogens are difficult to control once they are within the plant. Thus, sanitation is extremely important for disease prevention.

Avoiding leaf wetness

Environmental control of stem, leaf and flower diseases primarily involves preventing situations that allow moisture on leaves. In the first place, as mentioned above, splashing water can spread many pathogens' inoculum. Water drops on leaf, stem or flower allow many pathogens to infect host plants. The time the leaf must be wet will vary according to the organism and disease involved. For instance, with powdery mildew and bacteria, only a very short period (two hours or less) of leaf dampness is necessary in most cases. On the other hand, with Botrytis blights, downy mildews, or alternaria leaf spots, several hours of leaf wetness are needed to allow fungal spore infection.

Always try to avoid moisture on leaves. You can accomplish this in many ways. First of all, you may be able to alter your irrigation system to avoid overhead irrigation. Many greenhouses have largely eliminated foliage plant bacterial blight through this irrigation method change.

Never water a greenhouse crop late in the day. Late afternoon watering may lead to long periods of wetness at night. High humidity or dampness in a greenhouse (especially as temperatures drop toward nightfall) may cause dew or moisture condensation to form on crops. This, then, may allow disease development on stems, leaves and flowers.

Moisture also condenses on leaves if temperatures are allowed to drop too suddenly at other times during the day. For instance, in greenhouses with very efficient fan and pad cooling, if the vents and pads are opened too wide or too quickly mid-morning and a blast of cool air hits the warmer greenhouse air, a short, but sometimes damaging moisture condensation episode will occur! This can lead to powdery mildew infestations in many new or modern greenhouses. The popularity of double layer plastic covering has led to a problem with condensation drips producing moisture on leaves and flowers of many crops.

Overhead irrigation can lead to leaf wetness and spread pathogens.

As mentioned, it's especially important to control the humidity or dampness in a greenhouse at night. The best way to do this is to ventilate at the end of the day. Heat the incoming air slightly during ventilation. A little heating can dry humid air. Of course, this is more difficult in summer heat especially after using evaporative cooling pads for several hours! Good fan jets, horizontal air flow or air turbulator systems that move the air around the foliage within a plant's canopy will help control dampness in various spots within the greenhouse crop. Finally, proper crop spacing on the bench allows good air circulation around the leaves and prevents pockets of dampness in the leaf canopy.

Resistant plants

It's possible to take advantage of certain plants' resistance to some of the leaf, flower and stem diseases. Reliable, published data on this point is limited, but many growers have made good observations over the years about which plants are more prone to powdery mildew or rust or Botrytis. They have altered their cultivar selection lists accordingly. You can also manage the stem rot and wilt diseases, such as bacterial wilt and Fusarium stem rot through carefully observing and altering your plant cultivar selection list.

Fungicides

Many different pesticides or fungicides are used for leaf, stem and flower diseases in floral and foliage crops. Systemic pathogens, such as viruses and bacterial wilts, won't yield to any pesticide chemical treatment. Even localized bacterial blights and leaf spots are poorly controlled with pesticide sprays. Good bactericidal chemicals aren't available. The most common spray for bacterial blights and leaf spots is a mixture of copper (such as Kocide 101) and mancozeb. A new copper fungicide (Phyton 27) may prove more beneficial in the future. Streptomycin has been used as an antibiotic bactericide with variable effectiveness, but it may also cause phytotoxicity on greenhouse crops. We don't recommend streptomycin for wide use on greenhouse crops.

Powdery mildews can be controlled by a variety of products including Milban, Karathane, Cleary's 3336, Domain, Triforine, Sulfur, Rubigan, Banner or Bayleton. Products for Botrytis control have improved greatly in recent years and now include Daconil 2787, Chipco 26019, Ornalin, Cleary's 3336, Domain

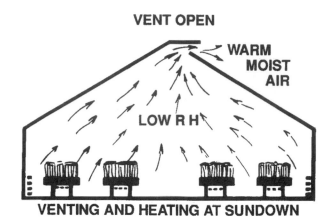

VENT OPEN

WARM
MOIST
AIR

LOW R H

VENTING AND HEATING AT SUNDOWN

Heating and ventilating to lower humidity.

or Exotherm Termil. For rust diseases, we generally turn to Manzate 200DF, Dithane DF, F-45 or M-45, Zyban or Bayleton. Other leaf spotting fungi require broad spectrum protectant type fungicides such as Dithane F-45 or M-45, Manzate 200 DF, Cleary's 3336, Domain, Daconil 2787 or Zyban.

Fungicide application procedures

Many growers don't achieve proper spray control of leaf, flower and stem diseases because they don't apply the products properly. "Wet down," or high volume sprays, continue to be the most effective method to get good performance from a chemical. Completely covering all plant surfaces that can be infected is necessary. Most importantly, this includes the leaf undersides. To achieve proper plant coverage use hydraulic applicators in a sweeping motion to produce reasonably fine droplets with good momentum. Use 200 to 400 gallons of dilute spray per acre depending on crop size and density.

Many growers are using low-volume applicators to save time, labor and chemicals. Many low volume methods are effective, but you must know precisely what you're doing and understand the biology of the particular pathogen/host system. Low volume application methods involve careful calibration and delivery techniques to avoid undue plant phytotoxicity. Certain products work better than others in a low volume applicator. More information on pesticide application appears in Chapter 22.

Managing root pathogens

Floral and foliage crops may have root rot diseases. Common root rot diseases include those caused by Pythium, Phytophthora, Rhizoctonia, Fusarium, Thielaviopsis and Cylindrocladium. These fungal pathogens may also cause crown rot and damping off diseases on floral and foliage crops.

With many diseases on many crops caused by many different organisms, the plan to manage root rots can seem quite complicated. As with leaf, stem and flower diseases, you can determine management strategies by thinking about

Good coverage, high volume fungicide spraying.

the "common denominators" of root rot diseases. Continue to think in an integrated and holistic manner.

Sanitation

Root rots generally result from soil-borne pathogens in the soil or growing media. A single, tiny grain of soil can be contaminated with many pathogenic fungi. Therefore, plan root rot management by thinking about eliminating (eradicating) pathogens by sanitizing planting media and media-associated items. For successful planting media sanitation, give equal attention to initial cleansing and recontamination prevention.

To understand the reason for growing media sanitation, begin with a bit of soil microbiology. To serve as a medium for plant growth, all soils must contain living microbes. Truly sterile soils or growing media won't support plant growth for very long, at least in a practical sense. Many beneficial animals and

Sanitizing growing media with steam.

Sanitizing growing media with an electric heating chamber.

plants (for example, fungi, bacteria, protozoans, nematodes) live in the soil. Some of the fungi, called mycorrhizae, even infect plants and work their good from within the plant roots. Thus, the concept of soil or growing media sterilization is faulty. But what about soil or growing media sanitation or pasteurization?

Harmful microbes can invade, live and spread in the soils or growing media. They cause root, stem and crown rot diseases on the plants if—and only if—two situations exist. Their numbers must be high enough and the particular environmental conditions favoring plant infection must exist for a fairly long time. Plant pathologists often refer to this as the soil inoculum potential.

Before placing plants in ground beds or using field soil or topsoil in potting mixes, sanitize with steam, steam-air mixtures, other types of heat or with fumigants to ensure that you don't start off with high numbers of harmful or "pathogenic" microbes in the soil or growing media. If you are using "soilless" or bagged media, sanitization prior to planting is usually not necessary. Most "soilless" media contain pathogenic microbes, but in small numbers. The aeration and drainage characteristics of "soilless" media aren't conducive to pathogen development. "Soilless" media are usually teeming with beneficial microbes. Some of these chemically inhibit or parasitize the pathogens. In other words, "soilless" media have low inoculum potential. Growing

Methyl bromide fumigation of growing media.

Sanitizing irrigation tubes.

media containing composted hardwood bark or pine bark are especially rich in inhibitory microbes. Sanitation methods can sometimes do these media more harm than good!

Soil can be sanitized with heat or chemicals. Heat methods are usually more foolproof and safer than chemical methods, but are also more difficult to do properly in greenhouses or nurseries without a steam source. For growing media in small batches, use an oven or other type of heating chamber. Spread the growing media in a shallow pan and slightly moisten it. Place it in a 200 F oven. Put a small potato (2 inches in diameter) on the surface of the growing media. When the potato's done, the growing media has been sufficiently heated to rid it of pathogenic microbes.

Many growers use steam under pressure to sanitize growing media. When using steam, 180 F for 30 minutes kills microbes, insects, nematodes and most weed seed. Tightly tarp the bed or batch of growing media and run steam delivery hoses under the tarp to the media. Steam sterilization's greatest shortcoming is that desired temperatures may not be reached throughout the soil or

A greenhouse water chlorinator.

growing media mass for the required time. The overall soil batch or bed may be hot enough except for one corner. The additional steaming necessary to heat the entire bed or batch tends to make the procedure expensive and time consuming. You may actually be oversteaming the majority of the soil mass. This can create an "overkill" situation, destroying beneficial microbes. Monitoring the temperature will help. Keep a long-stemmed thermometer handy and use it. Make sure that the media to be steamed is evenly but not overly moistened before treatment. Make no assumptions about the thoroughness of the job.

Many growers use electric soil pasteurizers to sanitize batches of growing media. This is another way of using heat economically and effectively. Because the process is time-consuming, most growers electrically heat the growing media at night. Using electricity at night also saves energy expense. Most devices are sufficiently insulated to hold heat long enough once it reaches 180 F or 190 F. Thus, a simple thermostatic cut off and a gradual cooling down period are effective. Electric soil pasteurizers come in various sizes, up to two yards capacity. Fill them loosely with slightly moistened soil or growing media, put the lid on tightly and begin sanitizing.

Many professional growers who sanitize growing media have turned to fumigation as an energy-efficient sanitation method. Methyl bromide (Brozone or Brom-o-gas) is a good, general soil-sanitizer. The procedure is not difficult but requires some planning since it can be dangerous. Methyl bromide is quite poisonous! Usually, fumigating piles of growing media relies on products available in small, 1 1/2-pound, pressurized containers filled with 98 percent methyl bromide and 2 percent chloropicrin. The chloropicrin is a potent tear gas and warns users of any leaks in the container, dispenser or soil cover. Use one of the various dispensers that are sold to puncture the cans and transport the liquid fumigant to a tray under the soil cover for evaporation. Once the pressure on the liquid is removed, the liquid boils at room temperature and the gas permeates the soil or growing media pores. Dispensers (available from your pesticide distributor) include Star Openers, the Jiffy Applicator and the Simplex Opener. Follow dispensing instructions carefully.

To prepare growing media for fumigation, proceed carefully as follows:

1. Loosen soil; make sure it's just moist enough for good seed germination and above 60 F.
2. Pile the media outdoors on a concrete floor or heavy plastic sheet. Methyl bromide is heavier than air and will settle beneath the pile if it's on a porous surface.
3. Don't pile the media more than 3 feet high. For piles between 18 inches and 3 feet high, use short pieces of drain tile to perforate with holes on 36-inch centers.
4. On top of the pile, place a tray or dish for the liquid to run into after the can is punctured.
5. Use 1 1/2 pounds methyl bromide per cubic yard of media. Use one, 1 1/2-pound methyl bromide container per tray.
6. Place unpunctured cans or tubes running from the dispenser into the tray and cover the pile with a gas-tight plastic sheet. Support the

Remove debris from benches between crops.

plastic cover a few inches above the pile so the gas will diffuse evenly over the pile before it starts settling in. Seal the tarp edges with extra growing media that won't be used. This will catch any fumigant that may leak out around the tarp edges.

7. From outside the gas-tight cover, puncture the cans. Have tape available to repair accidental or unnoticed punctures in the cover.
8. Let the fumigated pile stand undisturbed for 24 to 48 hours.
9. Remove the cover and let the media aerate for 72 hours. After one day, work or turn media to speed aeration. Little, if any, chloropicrin odor will be present. Most or all the methyl bromide should also be broken down.

You can use gas-tight vaults or drums for growing media treatment. Follow the above procedure, modifying where necessary. Be sure to pack the drums loosely so gas can permeate soil or growing media properly.

Soilless mixes may require sanitation if they become contaminated by contact with unsterile soil. Such contact occurs when you place mix on top of soil outdoors, when it contacts soil residue in a building or bin, or when you have used dirty equipment to mix it. Of course, a previously pasteurized or fumigated mix can likewise become recontaminated! Be careful how you handle and store "clean" growing media.

Sanitized growing media stored uncovered in areas where dust can settle on it can become contaminated. For example, growing media stored in flat filling machines can become contaminated when nearby transplanting areas are swept. The solution, of course, is to wash floors, keep media bins covered or sweep floors only after using dust prevention chemicals.

Any method you use to initially clean the planting media can be reversed by recontamination. If you use planting media as soon as it's sanitized, you run very little risk of recontamination. Many potted plant growers prefer to sanitize planting media in small batches just prior to use. If you have to sanitize larger

Good stress management begins with a controlled, stable environment.

quantities of planting media at any one time, it's important to devise a system where you don't have to move the planting media after sanitization. Each time planting media is moved, it could be recontaminated.

Sanitation also involves clean items associated with planting media. Many growers spend money and time to sanitize planting media, but then put the crop into dirty, reused pots! Sanitize pots, walkways, benches, hoses and water lines occasionally so they're free of fungal, root-rotting pathogens. The sanitizing agents on the market fall into several chemical groups. Many growers prefer to use phenolic sanitizers such as Amphyl. Others prefer sanitizers containing quaternary ammonium salts, such as Greenshield. If porous objects, such as clay pots or pieces of lumber, are to be sanitized, a sanitizing agent with some fumigation property is advantageous. For this purpose chlorine, as a 1:10 dilution of household bleach with water, or formalin, as a 1:18 dilution, are common and effective sanitizers. Remember that fumigant sanitizers lose their strength quickly after dilution. Use them immediately.

Pathogens causing root rot diseases often enter a greenhouse through contaminated watering ponds, pipes and other irrigation devices. If you have trouble with root rot pathogens constantly appearing in an otherwise sanitary greenhouse, investigate the water contamination levels. These tests are tricky and must be done by clinics. Water treatments using chlorine are easily done but wouldn't be prescribed generally unless indicated by a water analysis.

Using clean stock is another important sanitation aspect to rid the greenhouses of root rot disease. Although most crown and root rots aren't systemic in the plant's upper portion, they can be present there if they have been splashed up on muddy water droplets or moved up through careless handling. It's this surface contamination of cuttings that often leads to root and crown rots. Identify plants you intend to use as stock early in the production cycle. Treat these plants carefully and give them regular fungicidal sprays and drenches to ensure cuttings derived from them won't harbor root-rotting pathogens.

During the actual crop production, growers can do many things to help prevent and control root rot diseases through sanitation. Basically, all these activities involve keeping growing areas clean and tidy. Discard problem plants as soon as you notice them. They may not actually be infected with root rot, but they can lead to a start of root rot in the crop because they're weak and easily infected. Keep walkways clean and free of plant debris. Keep your feet off the benches. Avoid splashing water when watering crops. Keep hoses off the ground so they don't pick up contaminated muddy water from under benches or near walkways. Clean up benches between crops, carefully removing all plant debris. Clean out automatic "spaghetti tube" watering devices to destroy fungal pathogens between crops.

Environmental modification

Environmental modification varies according to the specific disease and pathogen involved in a root rot problem. One common denominator, however, is to avoid plant stress. Stress weakens a plant, making it susceptible to root rotting organisms. Avoid high salts (which may lead to Pythium or Fusarium root rot infection); wetness and poor planting media aeration (which may lead to Pythium and Phytophthora); dryness, especially between waterings (which may lead to Rhizoctonia or Fusarium); and cool temperatures (which favor some Pythium and Fusarium root rots). Compacted planting media, hit-and-miss fertilization programs and intermittent and uneven watering all indicate poor plant care that leads to root rot. Stress management means setting up a good growing environment and maintaining its stability.

One of the most crucial jobs in producing flower and foliage crops is watering the crop correctly! A crop that is over- or under-watered or watered irregularly can succumb to stress and then to infectious disease very quickly. Fertilization techniques also demand thorough irrigations at each watering to avoid excessive salt build up. The pH and mineral makeup of the irrigation water also can relate to stress. High calcium, alkaline ground waters, very common in many areas, can lead to stress and root health problems. Finally, soil moisture problems may result from other greenhouse situations. For instance, energy efficient greenhouses may have a water condensation and drip problem. Such a drip may produce an area of wet, waterlogged soil which will then stress part of the crop. If pathogens gain a foothold in these stress sites, they may rapidly proliferate, spread and infest the entire crop.

Another important stress management tool for managing greenhouse crop root rot diseases is to use a planting media with proper aeration and drainage. Planting media that holds about 40% (by weight) water when irrigated, but that has about 20% (by volume) air-filled pore space after irrigation and initial drainage generally promote root health.

Air-filled pore space is the volume of the growing media still filled with air after a thorough wetting followed by one day's normal drainage. This is figured as a percent of the growing media and root volume in the container. Roots need to take up oxygen and give off carbon dioxide if they are to function healthfully and properly. Air-filled pore space indicates how well the exchange of gasses between the root and the atmosphere occurs. Insufficient air-filled pore space increases root stress and the probability of root disease.

Growing media containing peat and composted tree bark will slowly age and decompose. As this happens, they become compacted. Eventually, the media no longer contains enough large pores or spaces to hold air, especially right after a thorough watering. Roots in these poorly aerated and waterlogged growing media will become sickly and rot. The plant will decline and eventually die.

Depending on the media, this can occur after years in an indoor garden or within weeks of planting. Many of the black peat (muck peat) growing media packaged and sold for home use are, in fact, poorly aerated because they are too compacted from the start. If a plant has become prone to waterlogging, you should conduct the tests mentioned below to see if compacted growing media is at fault.

Determining a growing medium's air-filled pore space (aeration)

To determine the percent of air-filled pore space, measure four quantities:

W1: The potted plant water-flooded weight, when all growing media pores are filled with water.

W2: The potted plant weight after draining for 24 hours.

W3: The container weight, emptied of growing media and roots and filled to the original growing media line with water.

W4: The empty container weight.

Procedure:

1. Flood the potted plant by slowly immersing it into a container of water until the water appears at the growing media surface. Immersing slowly avoids trapping air in growing media pores. Be careful not to let any water enter from the top of the container, or this may also trap air and give an erroneous reading.
2. Remove the potted plant and quickly transfer it to a bucket, being sure to catch any water that drains out. Record weight of potted plant, water, and bucket (W1).
3. Remove the potted plant from bucket and allow it to drain for 24 hours. Discard all drained water.
4. Re-weigh the drained potted plant and bucket (W2).
5. Mark the growing media line inside the container and remove the plant from the container. Place a watertight plastic bag into the container, fill with water to the growing media line mark, and weigh the water and the container (W3).
6. Empty the water from the container and record the weight of the empty container and plastic bag (W4).
7. Solve the following equation:

$$\frac{W1 - W2}{W3 - W4} \times 100 = \text{Percent air-filled pore space}$$

Resistant plants

The host part of the disease triangle is also worth investigating when trying to understand how to prevent and control floral and foliage crop root rot diseases. Some hosts are particularly prone to certain root rot diseases. For instance, some poinsettia cultivars are quite susceptible to Rhizoctonia. Some seedling geranium cultivars are prone to Pythium black leg. African violets and peperomia cultivars often have Phytophthora crown rot. Little published information is available about which hosts will be more or less resistant to which crown rot disease. Many growers, however, have made good observations over the years and have altered their plant lists accordingly.

Fungicides

Using soil drench fungicides to prevent and control many root rot diseases is common. Think of soil drench fungicides as a fourth line of defense. Effective fungicide use comes into play after considering resistant varieties, stress management and sanitation measures. To use soil fungicides effectively, prepare a proper fungicide dilution, then water or drench it thoroughly through the growing media.

Preventive programs usually involve monthly treatments. In this way, the fungicide acts as a protective barrier for the plant's entire root system. Soil fungicides may also act as effective curatives or eradicants for root infection. Inspect root masses frequently as part of a general preventive program.

Recently, many root-rot-managing fungicides have become available as granulars. Granulars generally work best when mixed into a planting media prior to potting, although soil surface applications also work, especially when followed by thorough irrigations. Granular products are best used as preventives.

Many good fungicides on the market control greenhouse crop root and crown rots. Sometimes combining products is necessary because many fungi are involved in these diseases. For this reason we often recommend a combination of Cleary's 3336 or Domain plus Banol, Cleary's 3336 or Domain plus Aliette,

Drench soil fungicides thoroughly into the growing media.

Cleary's 3336 or Domain plus Truban, or Cleary's 3336 or Domain plus Subdue. Terrazole can be used in place of Truban. Cleary's 3336 or Domain (benzimidazoles) are effective against Rhizoctonia, Fusarium and some other fungi. Banol, Truban, Aliette and Subdue are effective against Pythium and Phytophthora water molds. Subdue, Aliette and Banol, recently registered fungicides, are quite effective and long lasting. Banrot contains two active ingredients, so it can be used alone as a soil drench treatment.

Infectious disease prevention and control involve truly integrated programs in growing floral and foliage crops. Your efforts to manage diseases coincide with steps a good grower takes to produce a high quality crop ready for sale in the shortest time. Keep in mind the general management strategies outlined above and you will not have to be so concerned about specific infectious disease problems on a particular plant or crop.

The next chapter includes information on fungicides for floral and foliage crop disease control. This material follows this integrated control practices chapter because fungicides can't and won't ever perform properly and economically unless integrated into the larger health management program scheme outlined in this and previous chapters.

Chapter 6

Fungicides

Over the past several years, the types and numbers of fungicides to manage infectious diseases on flower and foliage crops produced in the greenhouse or outdoors have dramatically changed. The fungicide changes are the result of several stimuli in the industry. First is the universal increase in governmental actions. In the early 1970s new, sweeping pesticide legislation required that specific crop types, diseases and sites be listed on product labels for legal use. This triggered several years of research to "expand" the product labels already in use. Long lists of registered crop types appeared on products intended for greenhouse ornamental use. Fungicide chemistry also developed fungicides with new action modes; more are coming in the near future.

The more common floral and foliage crop diseases treated by chemical prevention or management are discussed below with reference to specific fungicides used in the United States today. The situation is constantly changing. Several diseases and many products probably won't be mentioned because of changes since this chapter was written! Their omission doesn't mean they aren't useful or as good as those that are mentioned.

Water mold root and crown rots

Phytophthora and Pythium fungi cause many root and crown diseases on floral and foliage crops. Twenty years ago, fenaminosulf (Lesan) fungicide became available. Although Lesan isn't as effective as many newer products, it was useful because it was labeled broadly for use on "flowers, shrubs and ornamentals in nursery beds." Lesan is no longer produced and sold in the United States, but is widely sold in other parts of the world as Bayer 5072.

Etridiazole (Truban, Terrazole) fungicide was the next new chemistry offered for water mold-incited diseases. This product continues to be widely used. Recently, manufacturers have developed several etridiazole formulations to better fit into greenhouse production methods. The product is available as a granular, wettable powder or emulsifiable concentrate. Etridiazole can cause plant stunting if used at too high a rate. This problem is more prevalent with highly organic growing media. Always use the lower of the labeled dosages ranges unless situations warrant more extreme treatments.

Etridiazole is also an active ingredient in Banrot fungicide. Banrot is a broad-spectrum product usable for other crown and root rots in addition to water mold diseases because it contains two active ingredients, etridiazole and thiophannate-methyl. More information about Banrot is provided below.

A few years ago, a product representing radically new chemistry became available. Metalaxyl (Subdue) is a fungicide with astonishing effectiveness on many crops at quite a low cost. Recently, a granular formulation was added to the 24% emulsifiable concentrate (EC). Growers can mix the 2% formulation into a potting media or plant bed prior to use. Metalaxyl is highly active. Only very small dosages, such as $1/2$ ounce per 100 gallons, are needed in many cases. Metalaxyl has been associated with plant stunting on rare occasions, probably because of accidental overdoses. Stunting will be worse in hot weather or when plants get excessively dry between waterings.

The future looks good for fungicides useful against water molds. Propamocarb (Banol) is an effective new product. Its current label requires relatively high rates, resulting in high cost. Fosetyl-Al (Aliette) is a new systemic for many flower and foliage crops. It can be sprayed on plants, where it will move to the roots and prevent water mold-caused diseases. Dosages are quite high for spray use, resulting in heavy spray residue. These last two products are quite safe on plant material.

Rhizoctonia, Fusarium, Thielaviopsis and Cylindrocladium damping off and root and stem rots

Benomyl (Benlate or Tersan 1991 DF) fungicide had long been used against these diseases but is no longer labeled on ornamentals in the United States. Similar to Benomyl, Thiophanate-methyl, as Cleary's 3336 or Domain, is still labeled on ornamentals. As flowables, 3336 or Domain can be used in injection lines but not with flowables or ECs used against water molds (see above).

Quintozene (PCNB or Terraclor) fungicide is also an older product. It, too, is effective against many "other" crown and root rotting fungi. Terraclor is labeled as a pot drench, as a soil mix and as a banded treatment for field grown flower and foliage crops.

Iprodione (Chipco 26019) provides some new chemistry against Rhizoctonia. Although effective, the crop list is limited and use is a bit tricky. Follow labeled instructions and don't use iprodione on plants not on the label.

Banrot, as mentioned above, contains two active ingredients. One is etridiazole for water molds. The other is thiophanate-methyl for Rhizoctonia and other root rots. This combination gives a broader activity spectrum against soil borne pathogens. The crop list is good. The company making Banrot produces an 8% granular formulation that can be pre-mixed into potting media as well as a wettable powder to be used as a watering-in drench.

Powdery mildews

Much new chemistry has recently emerged in the battle against powdery mildews and rusts. Thiophanate-methyl has been effective against powdery mildews. It shouldn't be relied upon solely, however, because the mildew fungi have been shown to mutate and become resistant to thiophanate-methyl under constant and sole use. The newer products, such as triadimefon (Bayleton, Strike), fenarimol (Rubigon), triforine, piperalin (Pipron), propicanazole (Banner) or dodemorph (Milban), have completely replaced benomyl in many cases. Many of these new products have long residual and good systemic action. They are extremely effective. Some, unfortunately, have a tendency to inhibit gibberellic acid synthesis and can cause plant stunting. Others are quite volatile and dissipate into the air too quickly when used outdoors, in hot weather or under windy conditions. The best bet when combating a powdery mildew problem with routine sprays is to change products every three or four applications. The future of fungicides for powdery mildews is changing rapidly. Labels are being "expanded" and new products are becoming available.

Rusts

Rust diseases on flower and foliage crops' leaves and stems are fought with older, topical protectants and newer systemic products. Of the older products, mancozeb (Dithane DF, M-45 or F-45 and Manzate 200 DF) and chlorothalonil (Daconil 2787) are useful if sprayed repeatedly and thoroughly. Oxycarboxin (Plantvax) is a systemic quite specific for rust disease control. Labeled for a few flower crops, it's a specialized product. Some newer systemics mentioned above for powdery mildews are also effective and labeled for rust diseases. Triforine and triadimefon (Bayleton, Strike) are particularly effective and well labeled for this use.

Botrytis blights

Botrytis blights have bothered flower and foliage crops for many years. As a result, 15 fungicides are currently registered for Botrytis diseases on greenhouse crops. Of the older products, benomyl and the other benzimidazoles were widely used. Botrytis in many places is now resistant to the benzimidazoles. The mancozeb products and chlorothalonil (mentioned above under rusts) are also very good.

The new, most widely labeled products for Botrytis diseases are vinclozolin (Ornalin) and iprodione (Chipco 26029). Ornalin is labeled as a spray, flower dip or bulb dip. Now available as a flowable, it's labeled for use in thermal foggers in greenhouses. Exotherm Termil, a chlorothalonil smoke formulation, is also used on many crops. All these products are quite effective if used according to labeled directions.

Miscellaneous leaf spots and blights

Many other flower and foliage crop leaf spots and blights occur occasionally. Although not widely prevalent, they can be devastating! Sixteen fungicides have at least one labeled use against these miscellaneous diseases. The older fungicides tend to have a broader activity spectrum against many leaf-spotting fungi. Unfortunately, few companies are seeking to "expand" these product labels. The three most "broadly labeled" useful products are chlorothalonil (Daconil 2787), the mancozebs (Dithane DF, M-45 or F-45 and Manzate 200 DF) and the combination product, Zyban. Zyban is particularly interesting because it combines the systemic thiophannate-methyl and the widely used topical fungicide, mancozeb.

Fungicides used to combat floral and foliage crop diseases in the United States

Disease	Common name	Brand name	Formulation
Bacterial diseases	Streptomycin	Agristep	21.2WP
		Agrimycin	21.2WP
	Fixed coppers:		
	Bordeaux mixture	Bordeaux mixture	12.75WP
		BordoMix	12.75WP
	Cupric hydroxide	Kocide 101	77WP
	Copper pentahydrate	Phyton 27	5.5EC
Botrytis blights, spots and stem rots	Bordeaux mixture	Bordeaux mixture	12.75WP
		Bordo-Mix	12.75WP
	Captan	Captan	50WP
		Orthocide	50WP
	Chlorothalonil	Exotherm Termil	20 Fumigant
		Daconil 2787	75WP, 4F
	Copper pentahydrate	Phyton 27	5.5EC
	Cupric hydroxide	Kocide 101	77WP
	Dicloran	Botran	75WP
	Ferbam	Ferbam	76WP
	Iprodione	Chipco 26019	50WP
	Mancozeb	FORE	80WP
		Dithane M-45	80WP
		Dithane F-45	4F
		Dithane DF	75DF
		Manzate 200 DF	75DF
	Thiophanate-methyl	Cleary's 3336	4F, 50WP
		Domain Fl	
	Thiophanate-methyl plus mancozeb	Zyban	75WP
		Duosan	75WP
	Vinclozolin	Ornalin	50WP
Botrytis damping-off and cutting rot	Thiophanate-methyl	Cleary's 3336	50WP, 4F
		Domain Fl	
Leaf and flower spots and blights caused by other fungi	Bordeaux mixture	Bordeaux mixture	12.75WP
		Bordo-Mix	12.75WP
	Captan	Captan	50WP
		Orthocide	80WP

	Chlorothalonil	Daconil 2787	75WP, 4F
	Copper pentahydrate	Phyton 27	5.5EC
	Cupric hydroxide	Kocide 101	77WP
	Ferbam	Ferbam	76WP
	Folpet	Phaltan	75WP, 50WP
	Iprodione	Chipco 26019	50WP
	Mancozeb	FORE	80WP
		Dithane M-45	80WP
		Dithane F-45	4F
		Dithane DF	75DF
		Manzate 200 DF	75DF
	Quintozene	Terraclor	2EC, 75WP
	Thiophanate-methyl	Cleary's 3336	4F
		Domain Fl	
	Thiophanate-methyl	Zyban	75WP
	plus mancozeb	Duosan	75WP
	Triadimefon	Bayleton TOF	25DF
		Strike	25DF
	Vinclozolin	Ornalin	50WP
	Zineb	Dithane Z-78	75WP
Leaf blights caused by foliar nematodes	Demeton	Systox	2EC
	Oxamyl	Vydate L	24L
Powdery mildews	Cyclohexamide	Acti-dione PM	.027WP
	Dinocap	Karathane	19.5WP
	Dodemorph	Milban	39EC
	Fenarimol	Rubigan	12.5EC
	Sulphur	Flotox	90Dust or WP
	Thiophanate-methyl	Cleary's 3336	50WP, 4F
		Domain Fl	
	Thiophanate-methyl	Zyban	75WP
	plus mancozeb	Duosan	75WP
	Triadimefon	Bayleton TOF	25DF
		Strike	25DF
	Triforine	Triforine	18.2EC
Rhizoctonia damping-off, root and crown rot	Ethazol plus thiophanate-methyl	Banrot	40WP, 8G
	Iprodione	Chipco 26019	50WP
	Quintozene	Terraclor	75WP, 10WP
	Thiophanate-methyl	Cleary's 3336	75WP, 4F
		Domain Fl	
Root diseases caused by soil nematodes	Fenamiphos	Nemacur	10G
	Fensulfothion	Dasanit	15G
	Oxamyl	Vydate-L	24L
		Oxamyl G	10G

Rots caused by other fungi	Captan	Captan	50WP
		Orthocide	50WP, 5Dust, 7.5Dust, 10Dust
	Dicloran	Botran	75WP
	Ethazol plus thiophanate-methyl	Banrot	40WP
	Quintozene	Terraclor	75WP, 10G
	Thiabendazole	Mertect 160	60WP
	Thiophanate-methyl	Cleary's 3336	4F, 50WP
		Domain Fl	
Rusts	Chlorothalonil	Daconil 2787	75WP, 4F
	Ferbam	Ferbam	76WP
	Mancozeb	FORE	80WP
		Dithane M-45	80WP
		Dithane F-45	4F
		Dithane DF	75DF
		Manzate 200 DF	75DF
	Oxycarboxin	Plantvax	5L, 75WP
	Triadimefon	Bayleton TOF	25DF
		Strike	25DF
	Triforine	Triforine	18EC
	Zineb	Dithane Z-78	75WP
Water mold damping-off, root and crown rot (Pythium and Phytophthora)	Captan	Captan	50WP
		Orthocide	50WP, 10Dust, 7.5Dust, 5Dust
	Ethazol	Terrazole	25EC, 35WP, 5G
		Truban	5G, 30WP, 25EC
	Ethazol plus thiophanate-methyl	Banrot	40WP, 8G
	Metalaxyl	Subdue	2E, 2G
	Propamocarb	Banol	66.5EC

This list is presented for information only. It doesn't endorse products mentioned, nor criticize products not mentioned. Before purchasing and using any pesticide, check all labels for registered use, rates and application frequency.

Chapter 7

Powdery Mildews

*P*owdery mildews attack many plants. While powdery mildew fungal infection seldom causes plant death, it almost always reduces the plant's looks and value. Serious economic losses are common.

Symptoms

Powdery mildews are easily recognized: white, powdery appearance on infected host portions.

Powdery mildew results from abundant, superficial hyphae (fungal threads) bearing many colorless conidia (spores). The pathogens usually attack young foliage, stems, flowers or fruit. On some hosts older leaves may be more susceptible than young foliage. Infection is frequently confined to upper leaf surfaces. Some powdery mildews attack only the lower leaf surface. Others show little preference for specific host tissues. Severe infection may result in yellow and curling leaves, general growth stunting, flower or fruit distortion and tissue death.

Sometimes powdery mildews cause various other symptoms, such as defoliation, witches brooms, drying and withering plant parts, leaf roll and virus-like symptoms. The pathogens don't directly attack plant parts below ground, but may indirectly affect root health. Severely infected plants frequently recover slowly even after the disease has been arrested. They may be more susceptible to insect attack, air pollution or root rot.

Disease cycle

The powdery mildew fungi undergo a rather simple life cycle on most floral and foliage plants. Simple, single-celled conidia (spores) form on fungal growth stalks called conidiophores.

A conidiophore begins as a swelling of the hyphae (the fungal thread growing on the host tissue). The conidiophore forms with a "generative cell" at its upper end. This "generative cell" continues to divide and produce conidia until

Powdery mildew on rose.

environmental conditions become unfavorable. In most cases, one conidia is produced each day on each conidiophore, on a somewhat diurnal or daily cycle.

Conidia usually mature 24 hours after formation and are then released from the conidiophore. Contributing factors to the release of spores into the air include a rapid drop in relative humidity, moving air, moderate temperatures and the sun's heating and drying effects on the fungal tissue. Although airborne, powdery mildew conidia aren't capable of long-distance dispersal.

Conidia usually germinate by developing germ tubes from the spore "corners." Various physical and chemical leaf factors affect the spore germination process. Following spore germination, the fungus penetrates the host epidermis and forms haustoria (nutrient gathering pegs or sacs) inside the epidermal cells.

Host cuticle and cell wall thickness influence the penetration process of many powdery mildew fungi. Since environmental factors such as light, temperature and relative humidity may significantly alter cuticle thickness, cuticle may vary considerably from plant to plant within a species or from leaf to leaf on the same plant. This results in plants' varying susceptibilities to infection and disease.

Powdery mildews' developing hyphae (fungal threads) soon branch and spread over the host tissue. Within 48 hours after inoculation, new, mature conidia can be present in optimal host-pathogen situations under ideal environmental conditions.

Environmental factors

Powdery mildews' occurrence, distribution and severity are affected primarily by temperature, relative humidity (RH), light, leaf wetness and wind or drafts. Because the effects of temperature, RH, light and wind (air movement) are often interrelated, it's important that all factors be considered holistically, in an integrated fashion.

Wind

Greenhouse growers have long recognized drafts as a contributing factor to the occurrence and spread of powdery mildews. Infection frequently begins near doors or other openings and gradually spreads. In the field powdery mildew

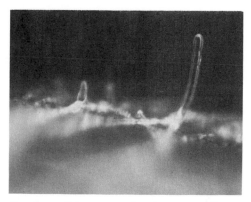

*Powdery mildew spores forming on a stalk or
conidiophore.*
Photo courtesy of American Phytopathological Society,
from *Compendium of Rose Diseases*.

appears on a plant and spreads in the direction of prevailing winds, apparently
caused by the movement of conidia from previously infected plant tissue.

Temperature
Response to temperature appears to be conditioned somewhat by the climate
in which the powdery mildew species has become adapted. For instance, optimum temperature for *Erysiphe cichoracearum* from cantaloupe is about 82 F
in collections from California's hot Imperial Valley. *E. cichoracearum* collected from squash in California's coolest agricultural region had an optimum
temperature of only 59 F. It seems most hosts can tolerate more heat than can
the powdery mildews that infect them. In the Midwest, powdery mildew diseases commonly develop in spring and fall. During the hot summer, many
plants recover from the disease almost completely. Few powdery mildews do
well at temperatures above 86 F.

Relative humidity
Although most powdery mildew spores germinate best in a saturated or nearly
saturated atmosphere, some germination occurs at all relative humidity levels.
The powdery mildews fit into three spore-germinating groups according to
their dampness responses:

1. Powdery mildews that germinate only under damp conditions
   (*Sphaerotheca pannosa*);
2. Powdery mildews that germinate optimally under damp conditions
   with a small percentage germinating under dry conditions (*E. cichoracearum, E. graminis, L. taurica* and *S. macularis*);
3. Powdery mildews that germinate well under conditions ranging from
   relative humidities slightly below saturation to very dry conditions (*E.
   polygoni* and *U. necator*).

Vapor pressure deficit (VPD) has been used as an indicator to predict powdery
mildew conidia germinability. Calculate vapor pressure deficit using the formula VPD=(1-RH)E; E is vapor pressure of water at saturation at a given

Wet, greenhouse rose leaves early in the morning.

temperature and RH is relative humidity. VPD-based models work only when environmental factors other than relative humidity don't limit the pathogen.

Leaf wetness

Recently, leaf wetness episodes on greenhouse crops have been linked to powdery mildew disease development.

Partial contact with small water droplets or water films benefit powdery mildew spore germination. Such conditions may also favor mycelial development and conidial production.

The relative humidity and temperature effects discussed above may actually influence pathogen development by causing leaf wetness. Minute moisture films on greenhouse crop leaves are visible diurnally (on a daily cycle) and can last several hours. Factors such as leaf-to-air temperature gradients, moving air, RH, incoming solar radiation, outgoing radiant energy loss and leaf transpiration undoubtedly govern the occurrence and persistence of leaf wetness.

Growers have used simple methods to control powdery mildew resulting from water on leaves. Soapy water helps to control peach powdery mildew. Rose growers spray plants with water at midday to reduce infection.

Heavy water films, which totally immerse the spores, reduce their germination but partial contact with small water films or floating spores on the water surface encourages conidial germination.

Light

Powdery mildews' diurnal spore maturation cycle may result from light variations. High light conditions, however, frequently include high temperatures resulting from incoming radiant energy. This incoming energy also tends to keep leaves dry. As a result, the pathogen growth may be more influenced by a temperature or leaf dryness response. Plant parts directly exposed to the sun frequently develop temperatures above the optimum for powdery mildew growth or even survival. Many powdery mildew infections develop on shaded foliage or on leaves deep within the canopy. This is especially true where a crop is shaded early in the afternoon, the hottest part of the day.

Wetting roses at midday to control powdery mildew.

Host nutrition and pH

Vigorous plants may be more susceptible to powdery mildew infection than plants suffering under various nutritional stresses because the vigorous plant's rapid growth may have created many leaves with thin cuticles. Except for nitrogen, soil amendments or soil pH alterations aren't practical ways to reduce powdery mildew infection. High nitrogen tends to result in soft, leggy plant growth, susceptible to many powdery mildew diseases.

Control

Control with chemicals

Growers have used sulfur to control powdery mildew for over 100 years. Sulfur is very effective for those plants that tolerate it. Sulfur-containing fungicides reduce sulfur's phytotoxicity and improve effectiveness. Such fungicides are still available, but have largely given way to new, more effective products.

Sulfur and sulfur-containing fungicides have been applied to plants primarily as sprays, but also have been widely used as dusts. In European greenhouses, growers first used sulfur in vapor form almost 100 years ago. They noticed sulfur was a particularly effective rose powdery mildew control when sprinkled on moistened, hot water heating pipes in the greenhouse. Sulfur fumes continue to protect many greenhouse-grown ornamental plants when fumigation occurs nightly through steam pipes or specially constructed sulfur-vaporizing units hung within the crop.

The major complaint against sulfur in both greenhouse and field is its temperature sensitivity. Sulfur's phytotoxic effects at high temperatures are well known. Regulating the temperature or dose when fumigating sulfur is often

Sulfur on steam pipes.

difficult. Water-cooled, sulfur vaporizers are helpful in preventing damaging high temperatures.

Many fungicides have vapor action against powdery mildews. Studies have shown that greenhouse disease control using heat volatilization of several fungicides is feasible but not currently labelled on existing products.

In the 1940s, dinocap (Karathane) began to replace sulfur, particularly on sulfur-sensitive plants and in cooler growing areas where sulfur was frequently unsatisfactory. Dinocap provides good control for many powdery mildews on many hosts. Although dinocap isn't completely safe, it's less phytotoxic on most plants than sulfur and is more effective at lower temperatures. Dinocap is still occasionally used on floral and foliage crops in the United States, but may burn open flowers.

Quinomethionate (Morestan), introduced in 1962, is active when volatized into greenhouses or sprayed onto plants. For many powdery mildews it's more effective than dinocap, but causes phytotoxicity on certain ornamentals.

Piperalin (Pipron) is a foliar fungicide with protectant and eradicant properties. Registered for use on rose, dahlia, phlox, zinnia and chrysanthemum in the United States, piperalin is one of the few fungicides that eradicates powdery mildew without harming plant foliage. It is very specific for powdery mildews and has limited usefulness for many greenhouse growers. It can cause an adverse reaction on some rose cultivars.

The benzimidazole fungicides comprise an important, systemic chemical group to control floral and foliage plant disease. Cleary's 3336 and Domain are useful benzimidazoles. Fungal resistance to benzimidazole fungicides has caused widespread concern. For this reason, growers should not rely solely on them for continued powdery mildew sprays.

Pyridine and pyrimidine fungicides, available for the last 20 years, are very effective against powdery mildews. Parinol (Parnon) is registered for use on roses and zinnias, but isn't available in the United States.

High volume spraying for powdery mildew control.

Triforine is a highly effective, systemic fungicide to control powdery mildews. It also controls rose black spot, rose rust and several other floral and foliage plant diseases. Triforine, often combined with thiophanate-methyl, is a sterol-inhibiting fungicide that affects both spore germination and mycelial development.

Dodemorph (Milban) belongs to the fungicide morpholine group and is effective against powdery mildews. It provides both protectant and eradicant effects, but has little residual activity on plant surfaces. Tridemorph is a closely related compound that has been used in Europe and other parts of the world, but not in the United States.

Triadimefon (Bayleton or Strike), propiconazole (Banner) and fenarimol (Rubigan) are other sterol-inhibiting fungicides now being used extensively on floral and foliage plants against powdery mildews. Repeated applications inhibit gibberellic acid and cause stunting in some plant species. Registrations of more fungicides of this type for powdery mildew control on ornamental crops are expected soon in the United States.

Imazalil (Fungaflor) is a relatively new systemic fungicide from the N-substituted imidazole group, not yet available in the United States. It has a broad spectrum of activity against fungal pathogens, including powdery mildews, and appears to be particularly active against benzimidazole-resistant strains of plant pathogenic fungi. The vapor phase distributes readily in confined areas, making it a good candidate for an effective greenhouse disease control fumigant.

Lights directed on roses control powdery mildew by helping keep leaves dry.

Control through proper chemical application

Because powdery mildews generally create superficial growth on host plant tissue, chemical application methods that wet plant and fungal surfaces well have traditionally been the most effective chemical way to control powdery mildews. Such application methods are usually classed as high volume sprays.

High volume sprays are adjustable to optimize their effectiveness in controlling powdery mildews by covering all potentially infectible plant surfaces. Leaf undersurfaces, bud tissue and flower parts can be particularly troublesome. Using an appropriate spreader-sticker will help. Adjusting nozzle direction or movement patterns with hand-held nozzles may be necessary to increase chemical application effectiveness.

The selected interval between sprayings can greatly influence effectiveness of either a preventive or eradicative powdery mildew spray program. If systemics aren't being used, rapidly growing crops usually need more frequent spraying. Weather patterns influence the need for follow-up sprays. If using a newer fungicide with growth regulator side effects, take care not to apply them too frequently. Spraying under cool and cloudy weather conditions also increases these products' tendency to damage plants.

Many growers have used low-volume chemical application methods to successfully manage powdery mildew diseases. These methods are fast, effective at first indication of a disease outbreak and don't expose the crop or applicator to excessive pesticide. A potential disadvantage is localized overdose on a crop. Furthermore, these methods don't directly wet fungal structures, possibly omitting fungitoxic action in an eradicative program.

Control through sanitation

Many powdery mildews that attack ornamental plants overwinter only by dormant mycelium in infected buds. Pruning and destroying infected buds reduces overwintering inoculum and helps manage disease where appropriate. Of course, such a strategy isn't particularly useful for greenhouse crop producers who deal in continuous, year-long production.

For mildews with broader host ranges, such as *E. cichoracearum*, removing weed hosts from the cultured plant area provides some control. As mentioned earlier, most powdery mildew conidia travel only short distances from infected plants. Keeping greenhouse doors closed to eliminate incoming air is an effective control procedure. Also, avoiding drafts and winds blowing over infested plants provides some mildew control.

Because powdery mildews occur on most ornamentals, little effort is made to exclude mildew-infected plants from international trade. Detecting dormant mycelium in the buds of transported plant material is difficult without an extended quarantine period. This unrestricted plant movement has resulted in widespread occurrence and losses from powdery mildew-infested crops in recent years. In addition, the possibility of introducing more damaging pathogen races or isolates into new areas is a serious concern.

Avoiding wind drafts was already mentioned as a way to reduce disease. The converse, promotion of wind drafts or air movement, is also an important aspect of powdery mildew control. Air movement over host plant surfaces helps prevent localized areas of high RH and leaf wetness. In greenhouses, effective ventilation and fan systems, properly spacing plants on benches and in beds, and appropriate greenhouse design create beneficial air movement. Outdoors, increasing plant spacing and clearing away surrounding buildings and vegetation increases drying air movement over host plant surfaces.

In the greenhouse, vent trapped, indoor air outward and bring in and heat outdoor air as the air cools at days' end to reduce relative humidity and prevent water condensation on plant surfaces. Using computers to constantly monitor and control greenhouse environments has made such a practice even more practical. Rather than dry crop surfaces after they become damp, vent and heat early in the evening to prevent moisture condensation in the greenhouse. In addition, air exchanges must be pulsed or applied intermittently to the greenhouse with sufficient heat to cause air exchange. A constant venting and heating procedure at moderate temperatures often doesn't change atmospheric conditions quickly enough.

Night curtains, high intensity plant lighting, and radiant heat pipes above a crop all effectively reduce powdery mildew development. Apparently, such practices add to or preserve the plant tissue's heat energy by preventing radiant energy loss. This prevents plant surface cooling below the surrounding air temperature. The end result is no water condensation on susceptible plant surfaces. Again, such practices must be employed properly, as soon as crops cease to be solarly heated or as soon as the greenhouse air begins to cool.

Note how the practices outlined above depend on several interrelating environmental factors, primarily temperature and air moisture. Although most powdery mildews develop more readily under moderate temperatures (below 86 F), growers haven't used high temperatures to manage powdery mildew except when the situation occurs naturally. Most crop quality falls off rapidly under cultural temperatures above 86 F and heating a greenhouse to these temperatures is not cost-effective. Of course, such a program might be practical if carried out for only a short time corresponding to the pathogen's diurnal cycling mentioned earlier. Determining such effective times would depend on knowing which pathogen structures were most affected by high temperature. Such information isn't presently available.

Chapter 8

Rust Diseases

*T*he plant rusts are so named because these fungi's spores are yellow brown to dark brown. Produced in large numbers, they give affected areas a rusty appearance. Many plant rusts exist throughout the world and some can be economically important. In general, those affecting floral and foliage crops are not too damaging, although on occasion they may be serious enough to severely limit growth, cause extensive defoliation, or even kill plants or plant parts.

Many rusts have two separate and distinct phases in their life cycles. Some rust fungi may produce as many as five different and distinct spore types as they complete this cycle. Further complications arise when the two phases are found on two different and completely unrelated plant types. Such rusts, called heterocious rusts, may have one phase on an herbaceous plant and the other on a woody plant. The herbaceous host may be annual or perennial grasses, flowering plants or ferns. The woody plants may be conifers or deciduous broad-leaved plants. In contrast to the heteroecious rusts are autoecious rusts, whose spore stages are all found on one plant type.

The spores produced by a rust in its yearly life cycle vary greatly from one rust to another. If a plant rust has all five spore types, it's called a macrocyclic rust. In contrast are the microcyclic rusts, which produce from one to four spore stages. Microcyclic rusts are almost always autoecious, having only one host. Most rusts affecting greenhouse crops are microcyclic. In this regard they are rather like other common leaf spotting fungi in their life cycle and control.

Spore stages

Each spore stage serves a specific function and has a specific name. Those working with the rust fungi use Roman numerals when noting which spore stage occurs on which host. The following paragraphs include general descriptions of each spore stage. Also outlined are the appearance of the spore stage and associated symptoms, the spore stage names, their functions and the numbers used to designate them. If you encounter a rust on a particular ornamental and know which spore stages occur on the plant, you will be able to devise a management program by referring to this outline.

The uredospores

The uredospore (also called urediospore) is the most common spore stage of economically important flower and foliage crop rust diseases. These spores' release often gives infected plant parts a rusty appearance. The uredospores can immediately germinate and infect. This spore stage generally acts as a repeating or build-up stage on the host resulting in rust lesions increasing rapidly. In some microcyclic rusts, this may be the only spore formed. Leaf tissue near the uredia (spore-bearing lesions) and sometimes surrounding them is discolored and may be pale green, yellow or tan-to-brown. The spores often appear first on leaf undersides. The spore stage number is II.

The teliospores

The teliospores (teleutospores) are resting spores and aren't common on floral and foliage crops. They may be formed in the same areas as the uredospores or in areas around the uredospores. Usually produced in abundance as the plant begins to mature and senesce, teliospores are darker than the uredospores and frequently are black. In many rusts, teliospores germinate only after a winter rest period. Thus, this spore stage is often responsible for the survival of the pathogen between greenhouse crops. The teliospores don't infect plant tissues, but each teliospore produces four very small spores called basidiospores. Teliospores are stage number III.

The basidiospores

The basidiospores are wind-borne spores formed from teliospores. In the autoecious rusts, these spores infect the same host on which the teliospores were produced. In the heteroecious rusts (those requiring two different hosts to complete the cycle) these spores infect the other host in the cycle. The spore stage is number IV.

The pycnospores (pycniospores or spermatia)

These very small spores don't infect plants but instead act as a sexual stage to provide for genetic recombination. The pycnospores (spermatia) are borne in small structures called pycnia or spermagonia. These are usually formed on upper leaf surfaces. They are very small and may be conspicuous as small black dots in brightly colored yellow or orange circular areas. The number for this stage is 0.

The aeciospores

These very light colored (pale yellow to orange) spores are borne in specialized structures, usually on herbaceous plant leaf undersides. The structures bearing the spores are often cuplike and are called aecia. Tissues associated with aecia may be discolored and swollen. On some plants aecia appear as ruptures or cracks in stems and branches. In heteroecious rusts, this spore infects the alternate host. The aecia frequently are associated with the pycnia, either surrounding them or being formed on the leaf surface opposite the pycnia. The aecial stage is number I.

Rust causing leaf damage on fuchsia.

Disease development

On most plants, the rust fungi infect leaves, but some may infect stems or flowers. The spores are spread by water splash or wind. Usually, lower leaves are infected first. Rust infections sometimes kill an entire leaf or several leaves. Rarely do the rust fungi kill a plant, although sometimes infected seedlings die. Sometimes severely infected leaves drop from the plant, leading to serious plant weakening and other troubles. Rust cankers from stem infections can result in stem death, especially if the canker is active for more than one year.

The rust fungi need leaf surface moisture for the spores to germinate and infect a plant. Infection can take place in as short a time as 4 $^1/_2$ hours. Thus, these fungi can be and are important in areas where nightly dew formation may be long. Warm days and cool nights result in such dew-forming situations. Moreover, cool temperatures (50 to 75 F) favor spore-bearing lesion production for most rust diseases. In areas where summer temperatures rise well above 75 F, rust diseases are not readily expressed, but they do come back in the spring and fall seasons.

Many greenhouse crop rusts have a prolonged incubation period between infecting the host and the host's developing symptoms. Thus, they can be moved long distances unnoticed on infected material. The rust diseases express themselves first by producing spores, often on leaf undersurfaces. Leaf yellowing often appears on upper leaf surfaces in spots associated with these spore-bearing lesions. Spray programs begun at the first symptom may not bring immediate results, since most fungicides are good infection-preventers but will not eradicate previous infections.

Managing rust diseases

Controlling rust fungi isn't easy. Plant debris that may harbor some rusts' resting stage (telia; III) should be buried deeply or taken away. In the infection's early stages, removing infected leaves can be effective. When irrigating, avoid

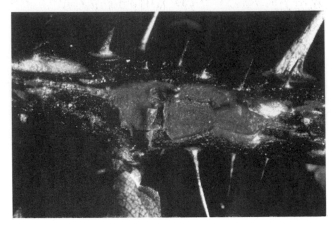

A rust stem canker on a rose.

getting the foliage wet, especially in the evening, for this favors rust fungi. Also avoid splashing water from leaf to leaf.

To avoid dew forming on the leaves or water dripping onto the leaves, vent and heat the greenhouse in the late afternoon. Venting and heating procedures are outlined in the previous chapter. Keep air circulating during this operation until the night temperature has been reached. Keep plants sufficiently spaced on the bench or in the growing bed to allow air to circulate freely about the leaves.

Chemical sprays can be effective if used correctly. Many fungicides act only as protectants, so apply them before infection. Good protectant control involves spraying at regular intervals. Because most germinating spores infect the lower leaf surfaces, direct fungicide beneath the leaf. Fungicides that protect against rust infection include mancozeb, triadimefon, triforine and chlorothalonil.

Some fungicides move systemically in the plant and may kill fungi once they become established. These fungicides include triadimefon, triforine and oxycarboxin. Unfortunately, not all are effective against all rusts. Many new ergosterol biosynthesis inhibiting (EBI) systemic fungicides are currently being developed. One newer one (propiconazole or Banner) seems to have broader activity against rust diseases.

In using chemicals, check labels to see what rust diseases they control and on what plants the materials can be used. Follow the directions for correct dose. The newer systemics can have growth regulator (stunting) and other phytotoxic side effects if used incorrectly.

Some plant varieties vary in their resistance to rust fungi. Both chrysanthemum and bearded iris varieties vary considerably in their susceptibility to rust.

Finally, growers should be aware that many rust diseases are spread via infected or contaminated plant material. Infection can occur on semi-dormant material or on seedlings' buds or tiny leaves. Many days or weeks can pass before the spore-bearing rust lesions are visible. If a rust disease appears to suddenly break out on a crop, it could be that infection occurred before the plant

Spacing plants to permit good air circulation will help control rust diseases.

material entered the greenhouse. In such cases, notify plant suppliers as soon as possible.

Greenhouse crop rusts

Because various rust diseases' life cycles are somewhat complex, an outline of some rusts found on greenhouse crops in the United States follows. Refer back to the various spore stage descriptions as you read about the rusts to understand the relationship between the life cycle and the diagnosis. Diagnosing a rust disease isn't difficult. Recognizing precisely which spore is present may be more difficult without prior knowledge. Even then, if there is a simple repeating cycle but no alternate host, manage the disease like any infectious fungal leaf spot for the most part. See Chapter 6 for appropriate fungicides to use on various host plants.

Ageratum

Puccinia conoclinii. Pycnia (0) are mainly on the upper leaf surfaces. Aecia (I) are on both surfaces and are in small groups. Uredia (II) are borne on the upper leaf surfaces and are cinnamon-brown. Telia (III) are chocolate-brown and found mainly on the leaf undersides.

Alcea (hollyhock)

Puccinia malvacearum. Pycnia (0), aecia (I) and uredia (II) aren't present. Telia (III) are small, raised, yellow-to-chestnut-brown mounds formed on the leaf undersides, on the petioles and on the main stems. The teliospores sometimes germinate in place and give the telia an ashen-gray appearance. This is found commonly on *Alcea rosea L.* wherever it is grown. Severe infections may kill young plants.

Anemone (windflower)

Puccinia anemones-virginiana. Pycnia (0) are unknown. Aecia (I) and uredia (II) aren't present. Telia (III) are compact, blackish brown structures on the leaf undersides. They may become ashen gray after teliospore germination.

Antirrhinum (snapdragon)

Puccinia antirrhini. Pycnia (0) and aecia (I) are not present. Uredia (II) are found on the older leaf undersides, although in severe cases all leaves may be infected. Stems and calyxes also may be infected. The rust pustules frequently occur in circles or concentric rings. The uredospores are chestnut brown. Telia (III) occur mainly on lower leaf surfaces and are blackish brown. This important snapdragon disease is general wherever *Antirrhinum majus L.* is grown.

Aster

Coleosporium asterum. Uredia (II) are round, orange yellow and are found on lower leaf surfaces. Telia (III) are reddish orange and also are found on lower leaf surfaces. This rust occurs throughout the United States and southern Canada but is more prevalent throughout the northern and western United States. Many aster species are infected. Pycnia (0) and aecia (I) are present on Pinaceae.

Centaurea (cornflower, bachelor's button)

Puccinia cyani. Pycnia (0) are scattered on both leaf surfaces. Aecia (I) occur in pustules on both surfaces and contain dark, cinnamon-brown spores. Uredia (II) probably are not present. Telia (III) occur on both leaf surfaces and are chestnut brown. The fungus may move systemically through the plant and the leaves may be covered with fungal spores.

Chrysanthemum (florist's chrysanthemum)

Puccinia horiana. Pycnia (0), aecia (I) and uredia (II) are not present. Telia (III) are mainly on leaf undersides, though they may occur on upper leaf surfaces and stems. The telia are pink at first, but as they mature, they become white. Because of this, the disease is called white rust. Teliospores may germinate in place to produce basidiospores, which are the means of fungal spread.

Puccinia chrysanthemi. Pycnia (0) and aecia (I) are unknown. Uredia (II) are found on leaf undersides on *Chrysanthemum* x *morifolium* Ramat throughout the United States and Canada. Uredospores are chestnut brown and frequently occur in rings. Yellow spots appear on upper leaf surfaces above infected areas. Telia (III) are not present in the United States.

Dianthus (carnation, pinks, sweet William)

Uromyces dianthi. Uredia (II) occur on both leaf surfaces and frequently form rings. They are dark cinnamon brown. Telia (III) are chestnut brown and also occur on both surfaces. Pycnia (0) and aecia (I) occur on *Euphorbia* spp. but not in North America.

Epilobium (fireweed, willow herb)

Pucciniastrum pustulatum. Uredia (II) are scattered or in small groups on discolored areas. They open by a central pore to release orange yellow spores. The telia (III) are small and develop into extended reddish brown crusts that turn blackish brown as they mature.

Fuchsia (fuchsia)

Pucciniastrum pustulatum. See Epilobium. Can be severe on *Fuchsia hybrida* Voss, especially those started in Oregon and Washington. Telia (III) are not present on *Fuchsia* spp.

Heuchera (coral bells)

Puccinia heucherae. Pycnia (0) are not present. Aecia (I) and uredia (II) are not produced. Telia (III) are formed mainly on leaf undersides and are chestnut or chocolate brown but often become ashen gray due to teliospore germination.

Impatiens

Puccinia argentea. Uredospores (II) are cinnamon brown and occur on both leaf surfaces. Telia (III) are dark chestnut brown and also occur on both leaf surfaces. Pycnia (0) and aecia (I) occur on *Adoxa moschatellina L. (Adoxaceae)*.

Iris

Puccinia iridis. Uredospores (II) occur on both sides of the leaves and are cinnamon-brown, frequently occurring in rings. Telia (III) also are found on both leaf surfaces and are chestnut brown.

Pelargonium (geranium)

Puccinia pelargonii-zonalis. Pycnia (0) and aecia (I) are not known. Uredospores (II) occur on lower leaf surfaces. They are scattered and frequently form irregular to regular concentric circles. Cinnamon brown, they are surrounded by the torn host epidermis. Teliospores (III) are pale brown and are mixed with the uredospores. They occur on *Pelargonium hortorum* Bailey throughout the United States.

Rosa (rose)

Phragmidium americanum. Pycnia (0) are not abundant and are inconspicuous. They are found on the upper leaf surfaces. The orange-yellow aecia (I) occur on the leaf undersides and on the petioles. Uredospores (II) are small and scattered on the lower leaf surfaces or on stem cankers. The telia (III) are on leaf undersides.

Salvia (sage)

Puccinia farinacea. Pycnia (0) occur on the upper leaf surfaces. Aecia (I) are cup-shaped and are present mainly on the lower leaf surfaces. Uredospores (II) are cinnamon brown and found on leaf undersides. Teliospores (III) are chocolate brown and also are found on leaf undersides.

Vinca major

Puccinia vincae. Pycnia (0) are scattered among aecia (I) on leaf undersides. Aecia are cinnamon brown. Uredospores (II) are dark cinnamon brown, scattered, and on leaf undersides. Teliospores (III) are chestnut brown and mainly on lower leaf surfaces.

Viola

Puccinia violae. Pycnia (0) form groups on both leaf surfaces. Aecia (I) are cup-shaped and occur mainly on leaf undersides. Uredospores (II) are cinnamon brown and found on leaf undersides. Teliospores (III) are chocolate brown and found on both leaf surfaces.

Chapter 9

Botrytis Blights

*T*he most commonly encountered flower and foliage plant blight diseases include those caused by Botrytis fungi, especially Botrytis forms grouped together as *B. cinerea*. Botrytis blights are most common in temperate zones where these pathogens attack a vast range of hosts including almost every herbaceous ornamental. In addition to ornamentals (herbaceous and woody), plant groups attacked by Botrytis are glasshouse and field vegetables, small fruits, bulb and corm-producing plants, monocotyledons and forest tree seedlings. *Botrytis* spp. are also spoilage organisms, causing considerable losses during storage and transit of flowers, fruits, cuttings, bulbs and greenery.

Botrytis symptoms include blights (of any plant part); leaf spots and stem cankers; corm, rhizome, root, tuber and seed rot; and damping-off of young seedlings. The disease name on a certain host may describe a characteristic symptom particular to the disease on that host. In general, all the diseases may also be called Botrytis blight. Thus, both cyclamen crown rot and exacum stem canker may be known as Botrytis blight. In each case, the causal agent is *B. cinerea*. Prolific gray or brown sporulation on blighted tissue characterizes most Botrytis diseases. The name gray mold, describing the fungal growth and sporulation, also describes some Botrytis diseases. Certain species cause diseases with more specific names, such as tulip fire caused by *B. tulipae* and narcissus smolder caused by *B. narcissicola*.

Pathogen cycle

Infection may result from germinating conidia penetrating undamaged tissue, stomata or wounds. It may also result from hyphae growing from a "food base" (either dead plant parts or extraneous organic matter) in contact with host tissue. *Botrytis* spp., especially *B. cinerea*, often occur as wound pathogens. Wound inoculations nearly always result in infection, but germ tubes can directly penetrate into undamaged tissue in some host-parasite relationships. Tulip leaves are infected by *B. tulipae* via stomata. Gladiolus leaves are infected by

Dormant hydrangeas blighted by Botrytis.

B. gladiolorum via stomata under low humidity, but direct penetration through the cuticle can occur under high humidity. Penetration into a host via stomata seems to be the exception rather than the rule in *Botrytis* spp.

Hyphal infections typically occur when infected flowers or leaves that act as food bases fall onto healthy tissue. Geranium petals that are readily colonized by *B. cinerea* provide such a food base when appressed to wet plant surfaces. Conidia seem less important than this mycelial inocula in establishing infections caused by *B. cinerea*, the most common greenhouse Botrytis that occurs on many hosts.

In a latent or dormant infection, the actual infection takes place as described above. Initial development in the host is macroscopically invisible. Further hyphae growth is delayed, usually for several days or weeks. When environmental conditions change, infections develop to a symptomatic stage. *Botrytis cinerea* produces such latent infections in flowers, such as cut roses. When infected flowers are initially packed, they appear healthy. After unpacking, they have large Botrytis spots!

Brown, moldy growth from Botrytis on a flower.

Gray, moldy growth from a Botrytis stem canker.

Most Botrytis conidia are dry and disperse in air currents in very large numbers. Sometimes they disperse in or on water droplets. Conidia in a film of water with dissolved nutrients for a minimum period (depending on temperature and other factors) is the basic infection requirement. Conidia can also be dispersed by insects such as bees and aphids.

To some extent, the mycelial "food base" inocula are independent of persistent water films containing nutrients. Water does improve adherence of diseased plant parts, such as blossoms, to healthy tissue. Mycelial inoculum size ranges from minute to very large. Wind-blown and rain-splashed plant debris containing mycelia are important dispersal propagules. Petals and whole senescent flowers, easily infected by *B. cinerea*, may drop onto healthy tissue or be wind- and rain-dispersed.

Conditions favoring disease development

Botrytis epidemics can happen very fast compared to most other diseases. The incubation and development periods can be very short. A conidium can form

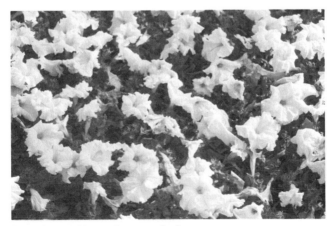

Botrytis attacking aging petunia flowers.

eight hours after infection. Many factors predispose crops to Botrytis diseases (see table), but other factors enable plants to escape infection. Growers can control epidemics by integrating various crop management practices.

Generally, Botrytis epidemics occur in cool, wet and humid weather, conditions that favor sporulation, infection and host stress. Surface wetness and temperature operate together to determine initial spore infection and subsequent development of lesions.

The host development stage is important with this disease. Typically, Botrytis diseases affect older, and especially senescent, tissues. In some crops, young tissues, such as bedding plant seedlings, are very susceptible. Between these two extremes is a normally resistant period when epidemics are rare.

Many factors may predispose plants to Botrytis diseases. Over-irrigating geranium cuttings, poor bulb crop drainage, crop shading and sheltering from drying wind and sun in landscape plant beds, plug seedling spacing, high greenhouse humidity and dense geranium stock plants are some examples of predisposing factors.

Some factors reported to definitely predispose plants to Botrytis disease epidemics

Predisposing factor	Host
Senescent tissue	Any rapidly drying floral part or any aging vegetative tissue
Frost	Any frost-damaged tissue
Low temperature	Gladiolus corms
Wounds, windblown sand, sun, hail, fungal lesions, rapid water intake	Most crops, especially grape
Crop fertilizers (deficiencies and excesses of nitrogen, phosphorus, potassium, magnesium, calcium)	Many crops, strawberry
Atmospheric pollutants such as ozone	Geranium, poinsettia
Ethylene	Cut flowers, especially carnation
Pesticides, fungicides and growth regulators	Tomato, grape
Insecticides	Snapdragon
Chloropropham	Tulip
Microorganisms, epiphytic organisms, parasitic micro-organisms, such as *Puccinia antirrhini*	Snapdragon
Corynebacterium zonale	Geranium
Mites	African violet

Air circulation is important for Botrytis control.

Disease control

Even though growers can control Botrytis diseases in many ways, they remain among the most economically destructive diseases, both in the greenhouse or field and in stored products. Controlling these fungi is difficult because they can attack crops at almost any growth stage and can infect all plant parts. *Botrytis cinerea* presents a constant threat of infection on many cultivated and wild host species and persists on dead plant material.

Botrytis spp. other than *B. cinerea* are more specialized pathogens with a restricted host range. Control is often possible with a single method such as

Flood or drip tube irrigation helps control Botrytis.

seed treatments or crop rotation. Fungicides are valuable, but only as protection. Since conidial infection can be rapid, infection may have occurred by the time the grower discovers unfavorable environmental conditions. Too, fungicide applications often can't be made when conditions favor infection.

In the early 1960s, systemic fungicides, such as benomyl, thiophanate-methyl and carbendazim, greatly enhanced chemical control. These chemicals were extremely toxic to *Botrytis* spp. and superior to the standard protectant fungicides such as captan and dichlofluanid. Disease control was significant, but developing *Botrytis* spp. tolerance to benzimidazole fungicides was a serious setback. This tolerance is widespread in the United States. As a result of resistance to benzimidazoles, growers turned to dicarboximide fungicides (iprodione, procymidone, vinclozolin) to control Botrytis diseases on many crops. Recently, dicarboximide-insensitive isolates of *B. cinerea* have been found. Chlorothalonil is also widely used and helps counter this resistance. Some newer ergosterol biosynthesis inhibiting fungicides may prove useful in the near future.

Crop-handling techniques are very important in controlling Botrytis storage diseases. Growers can control postharvest rots caused by *Botrytis* spp. with careful storage and shipping techniques: controlled temperature or controlled atmosphere storage, pasteurization, irradiation, fumigation and protecting fungicide dips.

Manipulating the greenhouse environment can promote conditions unfavorable for infection. Reduce the humidity by increasing temperature and venting at day's end. Air circulation over and through the crop also helps. In the greenhouse and field, plan cultivar selection and crop fertilization to avoid lush, soft growth as the weather cools, and choose cultural practices that include watering methods that don't leave foliage wet at night, spacing and removing weed hosts and debris to reduce inoculum.

Humid greenhouse conditions support Botrytis diseases. Growers can manipulate this environment to hinder disease development. In production, apply the following measures for disease control.

1. Before crop production, remove and destroy all diseased plants or plant parts and all plant debris and weeds. Continue sanitation practices throughout production.
2. Use disease-free seeds, plants, stock plants or propagating bulbs, corms, tubers, rhizomes.
3. Handle all plants carefully during transplant to avoid tissue damage.
4. Water plants without wetting foliage. Avoid overhead watering; it wets the leaves and jars the plants. In geranium crops it releases many spores into the air.
5. Space plants with leaves not touching so air circulates over all plant surfaces.
6. Keep humidity as low as practical. Evacuate warm moist air in the evening and replace with cool air. Heating this cool air reduces relative humidity and water condensation onto plant parts. Don't wet or wash benches or walkways in the afternoon or evening.

7. Apply protective fungicides thoroughly in anticipation of developing inoculum and susceptible plant growth in the greenhouse. Continue an appropriate application schedule as long as the weather remains cool and dry. With some crops, you may apply fungicides after blossoms open.

Field control can't be as easily manipulated as greenhouse control, but combining cultural methods and chemical controls applicable to specific crops is effective.

Chapter 10

Fungal Leafspots, Blights and Cankers

*I*n addition to previously-discussed pathogens that attack floral and foliage crops in greenhouses and outdoors, many other fungal microbes can—on occasion—cause serious diseases.

Diagnosis

All diseases discussed here are caused by fungi. Since various fungi are rather diverse, these disease symptoms are somewhat varied. Blights or spots may appear on leaves, stems or flowers depending on which disease is present. Many microbes in this group can attack all three plant parts. Few occur on below-ground plant portions. Generally, leaf blights and spots will be first visible on lower foliage or foliage within the canopy. Plants in the bench or bed center may become infected more readily. Try to find a fungal sign associated with the diseased plant tissue to determine the specific pathogen involved.

Most of these diseases are caused by fungi that produce abundant conidia (spores). Therefore, their specific diagnosis usually depends on finding the conidial-forming structures. Sometimes they are produced in great number, making infected tissue appear fuzzy. Snapdragon downy mildew is a good example. Sometimes they are produced in tiny, sac-like structures (pycnidia) or on thickened, fungal tissue pads (acervuli) at or just under the plant tissue surface. In these cases, you don't usually see the spores, but only the fungal tissue sacs or pads. These are often very dark and may occur in rings.

Some pathogens don't produce visible fungal structures or spores on infected plants. Naturally, this makes diagnosis harder. In such cases, prior disease knowledge or a plant clinic's help will be useful. Many pathogens produce rather characteristic spots or lesions. Alternaria, for instance, commonly produces a targeted or "bulls eye" spot of concentric rings. Cercospora often produces spots with a red border. Some cause a characteristic host reaction, such as rapid defoliation that occurs on roses shortly after a downy mildew epidemic begins.

Alternaria blight on zinnia.

Epidemiology

Although many of these diseases have specifically unique means of spread and conditions for development, important general epidemiological considerations also exist. As with so many ornamental diseases, the most common, long-distance spread of these fungal pathogens occurs when associated with diseased (but often symptomless) plant material or via crop debris. Debris from poorly cleaned seed lots often contains pathogen spores. Many spores have thickened, tough cell walls that allow them to transfer this way.

These pathogens spread locally in many ways. Splashing water from plant to plant is the most common local-spreading mechanism. Sometimes, spores spread in the air or on insect bodies. Handling diseased plants can also spread pathogens. Tiny bits of infected plant tissue can be carried on hands or tools to nearby plants.

Downy mildew growth on snapdragon.

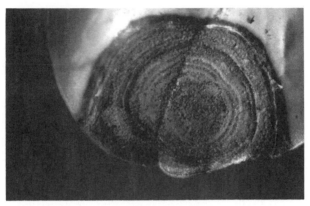

A fungal leaf spot with tiny, spore-bearing sacs visible as rings of bumps.

Water on the plant tissue surface is necessary for pathogen spores to germinate and for fungal strands to infect. Usually, water must be present for several hours. Temperature requirements for the various pathogens in this group vary. Most are warm weather pathogens, preferring damp night temperatures greater than 75 F and correspondingly hot, humid daytime weather.

Wounds are often necessary for infecting many pathogens in this group. Nutrients, primarily carbohydrates, are formed in the plant sap associated with these wounds. The pathogens respond to this nutrient. Many different wounds can serve as "infection courts." Tiny leaf wounds result from plant handling, spraying or watering, pruning wounds and insect-feeding or egg-laying wounds are all possible infection courts. Sometimes pathogens are limited to aging flower or leaf tissue. As such, they are noticeable but rarely serious.

Splashing water spreads fungal, leaf-spotting pathogens.

Sanitizing a walkway.

Control

Most of these diseases respond well to integrated control procedures carried out routinely or at the first sign of disease. In many cases, fungicide use isn't necessary. Control procedures are similar to those discussed in the chapters on powdery mildews, rusts and Botrytis.

At the end of a cropping cycle, clean up as much crop debris as you can. Sanitize benches and walkways with a quaternary ammonium chloride sanitizer to destroy pathogen spores or tissue in tiny bits of crop debris. For bed-grown plants, till the remaining residue into the bed so it can degrade to destroy most pathogen spores. Steam or fumigate the bed prior to reuse to kill remaining pathogens.

Inspect incoming plant material carefully. Discard plants or plant portions that show any disease signs. This procedure is often skipped in the rush to get a new crop potted or bedded out, but it is extremely important. Remember, this is the most common way these diseases gain entry to your production area.

Space plants adequately so that air can freely move over flowers and foliage. This is the most effective way to keep plant surfaces dry or to quickly dry them out after a dew period, overhead watering or rain. If possible, don't irrigate crops from overhead. This is especially important if you have noted disease in the crop or if that location has a history of disease. If you must water by wetting the plant leaves and flowers, do so early in the day. If plants are wet in the evening, they will remain wet all night. Finally, watch out for and eliminate drips onto plants under cover. Condensation drips from the inside surface of double layered plastics can be especially conducive to disease development.

Although sometimes serious, most of these diseases aren't considered "common." Thus, it's difficult to find fungicides specifically registered for their control. Chlorothalonil, thiophannate-methyl and mancozeb products are the most useful because of their rather extensive labeling. Thorough coverage sprays begun at the first disease sign will protect the plant from new infections. The sprays will not eradicate old lesions but may greatly decrease sporulation within them. Use high volume, wet-down sprays. If using wettable powders or dry flowable formulations, add a spreader-sticker to increase the fungicides' protective ability.

Several fungicide applications at intervals suggested on the label will be needed in most cases. Sometimes interval ranges (7 to 14 days) are given to guide you when crops are outdoors or subjected to watering that washes off protective fungicide residue. Obviously, if a lot of weathering occurred, use the shorter of the intervals. Keep an eye on residue so plants remain salable.

Fungal diseases of floral and foliage crop stems, leaves and flowers

Pathogen and disease	Important hosts	Comments
Alternaria blight	Carnation, geranium, pansy, schefflera, statice, zinnia, others	Causes ringed leafspotting, black or dark brown, and stem canker.
Anthracnose leafspots	Cyclamen, dieffen-bachia, English ivy, ficus, pansy, statice, others	Black spore-producing structures readily visible in lesions.
Ascochyta blight	Chrysanthemum, ferns, hydrangea	Can cause serious flower blight on mums, cutting rot and leafspot.
Black spot (Diplocarpon)	Rose	Begins on lower leaves; leaves turn yellow and drop. Very common on outdoor roses.
Brown leafspot (Coniothyrium)	Yucca	Small, target-ringed spotting, more severe on lower leaves.
Brown leafspot (Leptosphaeria)	Dieffenbachia	Circular spots, brown with orange halo or border.

Cercospora	Ficus, peperomia, statice	Usually causes a small, raised spot on leaf undersides. Often mistaken for oedema.
Conynespora leafspot	Aphelandra	Causes large lesion; a wound-invader.
Curvularia leafspot	Gladiolus, others	A warm weather disease.
Downy mildews	Cineraria, grape ivy, rose, snapdragon, others	A different pathogen on each host. Usually sporulates abundantly on lower leaf surfaces. Spores rarely seen on roses, where the disease causes severe defoliation.
Exobasidium leaf gall	Azalea	Causes thickened, red swellings on leaves. These later become covered with white spores.
Fairy ring spot (Cladosporium)	Carnation, cissus	A dark, powdery leafspotting; often ringed or targeted. Can be severe on carnations.
Fusarium leafspot	Dracaena	Reddish leafspots. Can be severe in warm climates.
Leafspot (Drechslera and Exerohilum)	Palms	Small, brown spots.
Leptosphaeria (Coniothyrium) canker	Rose	Causes cankering on canes. Cankers are slow to develop, but can eventually kill the cane.
Myrothecium leafspot	Dieffenbachia, gloxinia, peperomia, others	Fruiting bodies mature from cottony white to coal black. Can also cause crown rot.
Nectria (fusarium) canker or basal rot	Poinsettia, carnation	A stress pathogen. Rarely serious, but no known control once it appears.
Ovulinia flower blight	Azalea	A common galling and blighting disease of azaleas in the South.
Phyllosticta leafspot	Dracaena, English ivy, snapdragon	Brown, with rings.
Phytophthora blights	Petunia, vinca, others	Begins on lower portions of the plant. Can be severe in warm, wet weather.

Ramularia leafspot	Primula	Common on outdoor grown plants with overhead irrigation.
Rhizopus rot	Poinsettia, others	A general rotting of closely-spaced plant material in hot weather.
Sclerotinia flower blight	Camellia	Needs rain and warmth. Common in the South.
Septoria leafspot	Carnation, chrysanthemum, phlox, others	Can be serious in warm, wet weather.
Stem canker (Phomopsis)	Ficus	Occurs on weakened trees.

Fungicides registered to control miscellaneous fungal pathogens of ornamentals in the United States

Common name	Brand name	Pathogen	Hosts
Captan	Captan 50WP	Alternaria Diplocarpon Ovulinia Sclerotinia Septoria	Carnation Rose Azalea Camellia Chrysanthemum
Chloro-thalonil	Daconil 2787, 4F or 75WP	Alternaria Anthracnose Ascochyta Bipolaris Cephalosporium Cercospora Curvularia Cylindrocladium Dactylaria Didymellina Diplocarpon Fusarium Helminthosporium Ovulinia Septoria	Carnation, statice Statice Chrysanthemum, leatherleaf fern Parlor palm Syngonium Hydrangea, statice, leatherleaf fern Gladiolus Leatherleaf fern Philodendron Iris Rose Dracaena Maranta Azalea Chrysanthemum, hydrangea

Cupric hydroxide	Kocide 101, 77WP	Alternaria Anthracnose Cercospora Diplocarpon Septoria	Aralia, carnation Begonia Aralia, azalea, yucca Rose Chrysanthemum, yucca
Iprodione	Chipco 26019, 50WP	Alternaria Ascochyta Drechslera Fusarium Helminthosporium	Eighty-seven plant types listed on the label for this pathogen. Chrysanthemum Iris (see alternaria) (see alternaria)
Mancozeb	Dithane M-45, F-45, DF	Alternaria Anthracnose Ascochyta Cephalosporium Cercospora Curvularia Dactylaria Didymellina Diplocarpon Fusarium Leptosphaeria Mystrosporium Sclerotinia Sphaceloma	Aucuba, schefflera, zinnia Aucuba, anthurium, fatsia, hollyhock, pansy, Venus flytrap Chrysanthemum Syngonium Cordyline, ficus, hollyhock, hydrangea, peperomia, rhododendron, rose, statice Gladiolus Philodendron Iris Rose Dracaena, pleomele Dieffenbachia Iris Camellia Poinsettia
	Manzate 200DF	Alternaria, Anthracnose, Ascochyta, Cephalosporium, Cercospora, Coryneum, Curvularia, Cylindrocladium, Dactylaria, Didymellina, Diplocarpon, Fusarium, Fusicladium, Leptosphaeria,	(Manzate 200 is labeled on 121 of the listed ornamentals for all pathogens. See label before use.)

	Manzate 200DF	Melampsora, Melampsoridium, Monochaetia, Mycosphaerella, Peronospora, Pestalotia, Plasmopora, Phomopsis, Phyllosticta, Ramularia, Sphaceloma, Septoria, Sphaeropsis, Stemphylium, Taphrina	(Manzate 200 is labeled on 121 of the listed ornamentals for all pathogens. See label before use.)
Quintozene	Terraclor 75WP	Ovulinia Sclerotinia	Azalea Camellia
Thiophanate-methyl	Cleary's 3336 Domain Fl	Anthracnose Ascochyta Cercospora Didymellina Diplocarpon Entomosporium Ovulinia	Ornamentals Ornamentals Ornamentals Iris Rose Ornamentals Azalea, rhododendron
		Phomopsis Ramularia Sclerotinia	Ornamentals Ornamentals Azalea, rhododendron
		Septoria	Ornamentals
Triadimefon	Strike 25TOF	Cephalosporium Cercospora	Nephthytis Ageratum, marigold, phlox, rhododendron, zinnia
		Didymellina	Iris
Vinclozolin	Ornalin 50WP and 4F	Ciborinia Etromatinia Ovulinia Sclerotinia	Camellia Gladiolus Azalea Camellia, hyacinth, narcissus, scilla, snapdragon, zinnia

Chapter 11

Bacterial Diseases

*B*acterial diseases include galls, rots, cankers, wilts, fruit spots, leaf spots, blights and abnormal shoot formation. Agrobacterium, Corynebacterium, Erwinia, Pseudomonas and Xanthomonas are the most common bacteria that cause floral and foliage crop diseases. Some of these bacteria can live for several months or even years on plants, in soil or in water. They can live on or in roots, leaves, stems and flowers of host as well as non-host plants. They become pathogenic when environmental factors favor infection and disease development.

Pathogen cycles

Bacteria are "single-celled" organisms. They reproduce by simple cell division every 20 minutes, creating two, identical daughter cells. Bacterial populations on, in or near plants can quickly reach very high numbers. This development of high populations causes host plant damage.

Environmental factors influence bacterial population reproduction and death. As the temperature, food supply, moisture availability, pH, gas supplies and presence of toxic materials and antagonistic organisms changes, many bacterial cells die, but those that survive either adapt to the new environment (mutate) or were already better adapted (selection). In either case, these new types proliferate if the changing environment isn't too extreme. All bacterial cells usually don't die, even under extremely adverse conditions. The surviving cells often enter a state of reduced metabolism or dormancy until conditions again favor growth and reproduction. In such states they are undetectable on or in plants, pots, growing media or water. Bacterial cells undetectable within plants create difficult management problems in ornamentals.

Bacterial pathogens

Some plant pathogenic bacteria attack only one or a few plant species while others attack numerous plants, often in different families. Often you can't

Bacterial leaf spot on geranium.

determine the specific bacteria involved in a given disease solely on the basis of symptoms. More than one bacterium type can cause similar symptoms. Conversely, some bacteria can cause many different symptoms. The environment plays an important role in symptoms caused by most bacteria.

Erwinia chrysanthemi is the most destructive bacterium affecting foliage plants. It is a potentially destructive pathogen to many floral crops including chrysanthemum, carnation, poinsettia, begonia and African violet. It causes many symptoms, including soft rots, root rots, leaf spots, blights and wilts.

Erwinia carotovora causes soft rot symptoms similar to those caused by *E. chrysanthemi.* Beginning symptoms appear as areas of water-soaked tissue that later develop into a mushy, odiferous rot. Humid, hot conditions favor the disease. *Erwinia carotovora* causes soft rots in stems and basal crown tissue on many crops including poinsettia, dahlia, cyclamen, Calla lily and many foliage plants.

Pseudomonas solanacearum, the cause of southern bacterial wilt, can cause serious problems on plants in over 61 different families. Symptoms generally include wilting lower leaves followed by their yellowing and death. The leaf yellowing progresses up the plant as the infection develops. Often, the affected leaves will be only on one side of the plant. Infected plants' internal symptoms usually include a brown discoloration of the vascular system.

Pseudomonas cichorii causes brownish black lesions on the leaves of pothos, philodendron and several other foliage plants. It also causes a leaf spot and blight on chrysanthemum and geranium. Additional Pseudomonas species that infect herbaceous ornamentals don't normally have broad host ranges and sometimes attack only one species. The symptoms they cause range from leaf spots to blights, but leaf spots are more common.

Several *Xanthomonas campestris* variants cause leaf spots on many important flower and foliage crops including begonia, geranium, poinsettia, zinnia, dieffenbachia, philodendron and aglaonema. Symptoms range from small, well-defined spots to leaf death when infections are numerous. Yellowing leaf margins on some plant species occur frequently. Symptoms can vary depending on infected leaf tissue age.

Soft rot on poinsettia.
Photo by Robert E. Partyka, ChemLawn Services Corporation.

In addition to causing leaf spots, some Xanthomonas species can cause a systemic wilt and rot on geraniums, anthuriums, begonias, etc. Symptoms include leaf wilting and eventual death. Infected stems later turn dark green and collapse, killing the plants. These systemic Xanthomonads can go undetected in plants and cuttings during cool temperatures. Propagating and moving infected cuttings can result in tremendous losses under the warm, wet conditions that occur later in the season.

Agrobacterium tumefaciens causes crown gall in several floricultural crops including chrysanthemum, dahlia and rose. This pathogen also affects numerous woody and herbaceous perennials. Tumors or galls of varying size occur on infected plants' roots and shoots. Plants exhibiting these symptoms are unsalable and decline in vigor as they grow older.

Only a few Corynebacterium species are pathogenic on herbaceous ornamentals. *Corynebacterium fascians* causes a fasciation (shoot proliferation) from the geranium crown or lower stem and *C. poinsettiae* causes a canker or blight on poinsettia. Symptoms on poinsettias include leaf spots or blotches and longitudinal, water-soaked stem streaks. Both diseases are rare, but can cause serious losses.

Disease development

Plant pathogenic bacteria, unlike fungi, can't produce spores or other overwintering structures. Pathogenic bacteria are poor competitors with other organisms, especially in the soil. Surviving cells are likely to be very different in both morphological and physiological characteristics from actively metabolizing cells. Cells often survive in a state of reduced metabolism. They generally resist antagonistic factors. Cells gradually enter this state due to adverse environmental conditions, such as unavailable nutrients and/or a susceptible host. They survive best when they are in aggregates and when in a semi dehydrated state. Therefore, they most often survive in a semi dormant state in association with plant debris.

Many bacterial pathogens survive for one or more years in diseased plant tissues kept dry. As the debris becomes damp, nearby microorganisms multiply and decompose the tissue, bringing to an end the pathogen survival.

Bacteria stick to plant parts, especially leaves. These bacterial cells aren't easily removed, even by repeated washings with water. Most bacteria on exposed plant surfaces die, however, probably due to desiccation, exposure to sunlight, insufficient nutrients, antagonistic substances produced by other organisms and competition from other organisms.

Bacteria live within leaves as well as on leaf surfaces. Surviving bacteria are most likely to be found in aggregates in protected positions such as deep depressions between epidermal cells on leaf undersides because there is less drying sunlight there. Trichomes (leaf hairs) are survival sites for bacteria. Hydathodes and other natural openings also serve as protective positions.

In greenhouse production areas with controlled environmental conditions, survival pressures aren't normally as severe as those in the natural environment. Many pathogenic bacteria survive in plant debris for long periods in the warm, humid greenhouse environment. Natural environmental stresses such as drying and extreme temperatures reduce many bacterial pathogen populations that exist outdoors.

Some plant pathogenic bacteria can survive as saprophytes by growing on host plant leaves as harmless, surface-dwelling microbes. Later, when environmental conditions permit, they rapidly multiply, infest the plant and cause disease. Researchers have recently shown that several Pseudomonas, Xanthomonas and Erwinia species can survive in association with nonhost plants a long time. As environmental conditions permit bacterial growth, such surviving bacteria can resume metabolic activities. Rapid bacterial increases on leaf surfaces may occur during this time. These bacteria might act as inoculum sources since bacteria spread from leaf to leaf and plant to plant by splashing water and probably by insects; however, such a situation has yet to be officially documented.

It is common for bacterial pathogens to survive in association with seeds. They may survive within the seed or they may be surface contaminants. Drying kills many bacteria residing on external seed parts. Infested seed moved around the world is a common pathogen dispersal method. Many seed crops are grown in parts of the world where rainfall is minimal during harvest to help ensure freedom from seed contaminated with bacteria.

It's easy to see how a previously noninfested greenhouse site could be contaminated with symptomless plants harboring pathogenic bacterial populations either on the plants or in potting media. For this reason, obtain plants only from areas that enforce strict disease control measures. Potting media should never have crop debris in it, unless it has been thoroughly sanitized.

Humans are perhaps the most important agents in plant pathogenic bacteria dispersal. Simply touching an infested plant, then touching a noninfested plant, may result in pathogen transmission. Bacteria often are transmitted from plant to plant on the knife during propagation and pruning procedures. Bacteria surviving in plant debris and in soil are often carried from one area to another on shoes and machinery.

Water is an excellent bacteria dispersal agent. Many producers use overhead irrigation to water plants. In addition, pesticide application may disperse bacteria. Dipping cuttings or plant divisions into fungicide or rooting hormone dips spreads bacteria. When bacteria are present on a cutting, they can be released into the dip where they can later cling to other plants.

Insects may also play an important role in pathogenic bacteria dispersal and survival. This is probably of minor importance in properly maintained production areas, although greenhouse whiteflies can carry the geranium wilt pathogen to healthy plants.

A moisture film is usually required for plant pathogenic bacteria infection. Moisture also acts as an important vehicle for moving bacteria from a survival position to a natural opening or a wounded area. Moisture can affect the pathogen, the host or both. When moisture is available, plant tissues may be more succulent and provide a more conducive environment for bacterial multiplication after they enter the plant's intercellular spaces. It's essential that ornamental producers prevent moisture from remaining on leaf surfaces for extended time periods.

In general, low temperatures don't favor bacterial disease development. As temperatures increase, the bacteria increase their metabolic activities and cause disease if other factors are favorable. Disease development is usually most rapid as temperature increases, espccially as it rises to levels stressful to the host plant.

Nitrogen influences development of ornamental plant bacterial diseases, but in different ways depending on the disease. Several foliage plants tolerate more bacterial pathogens when fertilized with excess nitrogen. In some situations, however, more disease develops on plants fertilized with more nitrogen.

When all conditions are favorable, both saprophytic and pathogenic bacteria can invade a plant's intercellular spaces. Entrance occurs either through wounds or natural openings. Once inside the plant, the pathogenic bacteria generally multiply intercellularly. In some vascular diseases, however, they multiply in the xylem. Bacteria release biologically active substances that kill and disorganize tissues before invasion. As a result, cell substances are released into intercellular spaces and bacteria use these nutrients for growth and reproduction processes.

Bacteria's different effects on plants are a result of substances released from host cells. In many cases, very limited bacterial reproduction occurs within the plants. Systemic bacteria spread can still take place, however, resulting in the symptomless but infected situation mentioned earlier.

Control

Controlling ornamental bacterial diseases involves cultural and chemical prevention. Good sanitation practices such as destroying diseased plants and keeping production areas clean are essential, since many bacterial pathogens can survive on both host and nonhost plants near production areas. Pathogens also may survive in potting media and soil and come from shoes, shovels and hands.

Good cultural practices include proper fertilization, spacing plants to provide good air movement and prevent water splash and watering to avoid wet foliage. Plants stressed from improper fertilization, poor watering practices or inadequate lighting are usually more susceptible to diseases. Good watering practices are essential to reduce bacterial disease severity and spread. Plant wounds enhance and often are a prerequisite for disease development.

Pathogen-free propagative materials are essential as most bacterial pathogens can be introduced to a crop or a greenhouse on seeds and cuttings. Seeds produced in dry climates without overhead irrigation are more likely to be free of bacterial pathogens. Bacterial pathogens may be eliminated from seeds by soaking them in hot water (122 F), sodium hypochlorite or hydrochloric acid. Obtaining pathogen-free seed is absolutely necessary to ensure healthy plant production.

Cuttings obtained from culture-indexing programs will lessen the likelihood of bacterial pathogen introduction into a crop. Of course, the cuttings actually obtained in such programs aren't directly culture indexed. They merely came from grandparents or great-great-grandparents that were culture indexed in a laboratory. Nevertheless, companies offering such material have taken great care to see that the plant material remains free of bacterial infestation as the culture indexed generation is built up to production block quantity. The material is not resistant to infection, however, and can become infested at any time! Many growers don't realize this. They may buy culture-indexed material and place it next to their own, held-over material or mistreat it in other ways. The result is often an infestation of the cultured material and a disease epidemic.

Streptomycin, oxytetracycline and copper sprays control or reduce floral and foliage crop bacterial diseases. Exercise caution because these sprays may cause phytotoxicity on various plants under certain environmental conditions. Mancozeb mixed with fixed coppers is more effective in controlling certain bacterial diseases than fixed coppers alone. The mancozeb-copper combinations are most effective when mixed 90 minutes prior to application. Many bactericide studies have been conducted but, unfortunately, growers still haven't effectively controlled bacterial diseases with chemicals. A new product, 5.5% copper pentahydrate (Phyton-27) is a systemic that appears safe on many ornamentals. It may provide superior control of floral and foliage crop bacterial diseases in greenhouses or outdoors. A chelated bromine water additive (Agrobrom) used to control algae and slime has also been shown to reduce bacterial diseases when misted continuously over foliage.

What is a culture-indexed cutting?

Many bacterial, viral and fungal wilt pathogens are transported into a crop by an infested cutting. The cutting is normal and healthy, but contains tiny amounts of pathogen that later develop and spread disease throughout the crop. Growers reduce the likelihood of getting infested cutting material by purchasing "cultured" or "culture-indexed" cuttings.

Cultured cuttings are the end point in an elaborate program to develop and produce pathogen-free, vegetative propagative material. These programs, begun in the late 1940s, have preserved the quality and healthfulness of many important flower and foliage crops. Here's how the programs work.

Two or three years before final sale of a particular cultured cultivar, a plant (believed to be pathogen free) is selected. It may have been heat-treated or meristem-derived. Cuttings taken from this plant are brought directly to a laboratory. In the lab, the cutting's basal end is removed and subjected to various culturing tests for bacteria and fungi and to various indexing tests for viruses. This is the origin of the term "culture-indexing."

The top portion of these cuttings is carefully rooted and potted in an isolation greenhouse. If all tests are negative, the cutting is grown on to form a "nucleus" block to yield more cuttings. These, too, are run through the culture-indexing procedure to check for missed pathogens the first time around. Sometimes, a third round of culture-indexing is done.

Cuttings that survive all this become the "mother" block of the "mother block system" of plant increase. Cuttings taken from these mother block plants go into "increase" blocks. "Production" blocks are formed from increase blocks.

Growers who purchase "cultured" material get plants from these production blocks. Production block cuttings are not directly cultured, but are at least four generations from culturing. Furthermore, production blocks may be in different greenhouses--even in different countries-- from the original, nucleus material. The material may have been pathogen-free to start with, but could possibly be infested again when it arrives.

There's nothing really wrong with this proven, successful system. As long as growers buy cultured material from the company that administers the original steps in the program (or a licensed propagator), they will likely receive pathogen-free material. These companies have their reputations to uphold. They take great care to protect material properly as it is being increased to the production block stage.

Chapter 12

Wilt Diseases

Wilts are recognized easily in most plants. Common wilt symptoms include drooping or wilting of a leaf portion, an entire leaf, a plant portion or an entire plant. The wilt is often accompanied by a clearing and yellowing of leaf veins, a vascular browning and whole plant stunting. Wilt symptoms may occur suddenly or develop over time.

Wilt disease losses vary greatly. Slight wilting of one or several leaves on mature plants close to sale or harvest may not result in extensive losses. Severe wilting of young plants, however, often results in plant death.

Noninfectious wilt diseases

Wilting may result from noninfectious, environmental causes. Overly dry soil or potting mix permanently damages roots. If soil moisture stress isn't too great, plants wilt but recover when moist again. The plant won't recover from serious episodes of dryness.

Prolonged soil saturation or waterlogging may create an oxygen shortage in roots, and plants wilt. Recovery occurs when soil drains and is aerated, if the low oxygen stress hasn't been present for too long. If soil is waterlogged for prolonged periods, roots may die. Also, infectious root rotting may occur after a time. In either case, plants suffer permanent damage. Infectious root rot pathogens can thus cause wilting, even though they aren't normally considered wilt disease pathogens.

Plants also may wilt in response to physical root injuries. Chewing insects, burrowing animals or nematodes may feed on roots, reduce root volume or root function and cause wilting. Physical evidence of root damage caused by these pests is usually visible in the roots. High soluble salts in soil or potting media can damage roots and result in wilting. This wilting becomes visible under high light and temperature conditions, when plant transpiration rates are especially high or when chemicals containing wetting agents are drenched into the planting media.

Chrysanthemum Verticillium wilt.

Infectious wilt diseases

Bacterial and fungal "wilt" pathogens are so named because the infection symptoms result from the pathogen's presence and activities in the plant's xylem or water-conducting tissues. Vascular wilt diseases may be caused by several fungi and bacteria.

The fungi known to cause ornamental vascular wilts include *Ceratocystis* spp., *Fusarium oxysporum*, *Verticillium albo-atrum*, *V. dahliae* and *Phialophora cinerescens*. Different host plants are attacked by special forms or "races" of *F. oxysporum*. Thus *F. oxysporum f.* sp. *chrysanthemi* attacks chrysanthemums and *F. oxysporum f.* sp. *dianthi* attacks carnations. *V. albo-atrum* and *V. dahliae* can attack a wide range of ornamental plants while *Phialophora cinerrescens* is primarily a problem on carnations.

Bacteria known to cause ornamental vascular wilts include *Erwinia chrysanthemi*, *E. caratovora f.* sp. *caratovora*, *Pseudomonas caryophylli*, *Pseudomonas solanacearum* and various pathovars (pv.) of *Xanthomonas campestris*.

Most fungal wilt pathogens are soil-borne. They attack primarily through plant roots. Root hairs, root tips or root wounds exude chemicals that serve as an initial nutrient source for potential pathogens.

Bacterial pathogens require a wound or natural opening to invade plant organs. Bacterial wilt pathogens usually invade leaf tissues. Both bacterial and fungal wilt pathogens may be carried in the stems of propagation cuttings and may be present in such cuttings without showing symptoms.

Diagnosis

Basically, vascular wilt pathogens cause wilting, a nonspecific symptom that doesn't help determine which pathogen is involved. Indeed, as noted earlier, it doesn't even indicate that an infectious disease is present. The specific plant parts affected and the wilt pattern progression may indicate the specific wilt disease pathogen present. The host plant involved may also indicate the pres-

Leaf yellowing is an early wilt disease symptom.

ence of one pathogen over another. For instance, when geraniums wilt, it's almost always because of *Xanthomonas campestris* pv. *pelargonii.*

Generally, wilt symptoms result when pathogenic organisms invade a host plant's xylem or water-carrying vessels. During this invasion, the fungi or bacteria interfere with water and nutrient translocation by clogging water-conducting vessels. Fungi-causing wilts remain almost entirely in the vascular tissues until plant death. Bacteria-causing wilts often spread into other leaf or stem tissues, causing a general leaf blighting or stem rotting and cankering before plants die.

Xylem vessels may be blocked by the pathogen's physical presence, by the pathogen's gummy exudate or as a result of pathogens rupturing plant cells. Toxins secreted by some wilt-inducing fungi, particularly *Fusarium* spp., also may be involved in symptom development.

Initial wilt disease symptoms are a yellowing of a leaf portion or perhaps several leaves on a plant side. The yellowing can occur on older or younger leaves depending on the pathogen involved. Leaf margins may wilt a few days before the entire leaf wilts and yellows. In severe cases branches and entire plants eventually yellow and wilt. Wilted leaves may turn brown, but usually remain on the stem.

In Verticillium wilt and Fusarium wilt, early wilting may involve only a leaf half or one plant side. For most wilts caused by these fungi, the symptoms begin first on older leaves and progress up the stem. Branch and branch tip defoliation and dieback may also occur in wilts caused by these fungi. Infected plants can be severely stunted.

Examining vascular tissue can help further diagnose the problem. A brown or reddish brown discoloration of the water-conducting or vascular tissues is often evident in plants attacked by Verticillium or Fusarium. This discoloration is the most reliable field symptom of a wilt pathogen infection. When a stem or branch is examined in cross section, an incomplete-to-complete brown ring is visible.

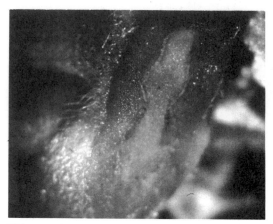

This vascular tissue browning is caused by wilt pathogens.

In the case of bacterial blight and wilt caused by *Xanthomonas campestris* pathovars, symptoms usually begin on individual leaves. Leaf margins wilt. Entire leaves yellow and wilt as the infection spreads to leaf petioles. When bacteria invade the stem water-conducting tissues, stem tissue turns brown to black. Externally, stems appear gray and dull, eventually rotting in cankered areas. Roots can be blackened by bacterial blight but not decayed. Bacterial blight symptoms may proceed up or down a stem.

Infection by the Southern bacterial wilt organism, *Pseudomonas solanacearum*, normally occurs through roots. Roots and then stems decay, turn brown and then turn black.

Wilting symptoms of carnation slow wilt (Erwinia) develop over several months before killing a plant. Symptom development in other Erwinia wilts may be much more rapid.

Although some cuttings infected by a wilt pathogen may express no recognizable symptoms during propagation, others may indicate a problem exists. Infected cuttings may sometimes fail to root, decay from the base upward, root slowly or develop typical wilt symptoms prior to transplant.

Pathogen survival and spread

Most vascular wilt pathogens can readily survive in soil or potting mixes. Generally, bacterial pathogens survive shorter periods (one to six months) than do *Fusarium* spp. and *Verticillium* spp. (several years).

In most cases, the pathogens, especially the bacteria, survive in plant debris. Bacterial survival time depends on the amount of pathogen in the tissue and the plant material decay rate. *Fusarium* spp. and *Verticillium* spp. survive in soil or plant debris as resting structures. The resting structures for Fusarium are called chlamydospores. Microsclerotia are the resting structures for Verticillium. They have thickened, outer cell walls and resist weathering and attack by

Advanced bacterial wilt and stem rot in geraniums.

other soil-dwelling organisms. The resting structures usually form in plant tissue shortly after the host tissue dies and begins to dry up.

Fungal and bacterial wilt pathogens are dispersed when mycelium, resting structures or cells are carried by water movement through growing media, by splashing of growing media, by physical movement of growing media in pots, flats or on equipment, in dust and in or on cuttings. Water movement, specifically water splashing from one plant to another during irrigation, is a primary cause of spreading bacterial wilt pathogens. Bacteria may spread by physical contact between leaves upon which a water film is present. In the case of geranium bacterial blight, the pathogen can also be spread by insects, especially whiteflies.

Infected stock plants may be symptomless and a grower may unknowingly spread the pathogen. Infected cutting knives are also an important means of spreading wilt pathogens. A cut through an infected stem may leave wilt fungi or bacteria particles on the blade surface. If the knife blade is not disinfested between cuts, an entire batch of cuttings may be contaminated. Infected cuttings are extremely important in the spread of wilt pathogens. Wilt disease can be spread throughout the world in this manner.

Using infected cuttings can lead to contaminated cutting beds. If infected cuttings aren't removed before they yellow and die, the cutting bed media may become infested with the wilt pathogens and serious wilt epidemics may result. This can spread pathogens into batches of healthy cuttings put into the beds at a later date.

Factors favoring disease development

Generally, high air and soil temperatures and high relative humidity favor bacterial wilt disease development. Symptom development of geranium bacterial blight is generally enhanced between 70 F and 86 F but is largely suppressed at 50 F to 59 F or at 90 F to 100 F. Symptom development for

Infested growing media, often as airborne dust, is the primary means of spread of wilt pathogens.

chrysanthemum bacterial blight can be slowed or arrested by temperatures less than 81 F with less than 80 percent relative humidity.

Southern bacterial wilt derives its name from the prevalence of high soil temperatures and high soil moisture in the southern United States, both of which favor development of *Pseudomonas solanacearum.*

Development of Fusarium wilt diseases also is affected by temperature. This may relate to host stress more than pathogen response. Development of carnation Fusarium wilt is favored by soil and air temperatures greater than 70 F. The temperature optimum for chrysanthemum Fusarium wilt is somewhat higher (81 F to 90 F). Chrysanthemums prefer warmer temperatures than carnations; thus, their heat stress thresholds are higher. Although wilting symptoms fail to develop at lower temperatures, infected plants may still be stunted.

Disinfestation of knives.

INDEX STEPS 1&2

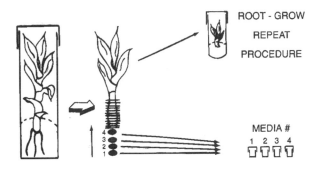

Culture indexing of plant material used as wilt free stock.

Mineral nutrition can affect the rate of wilt disease development. Nutrition below or above that optimum for plant growth will usually encourage disease development. Under high nitrogen conditions, geranium bacterial blight symptoms develop more rapidly than under low nitrogen conditions.

The type of nitrogen fertilizer may also be important. Chrysanthemum Fusarium wilt symptoms are less severe with nitrate nitrogen fertilization than with half ammonium, half nitrate nitrogen fertilization. Symptom severity also decreases with increased soil pH.

Generally, plants growing under optimum cultural conditions will be less susceptible to vascular wilt pathogens. Plants may be predisposed to wilt disease by stressful environments such as high or low air or soil temperatures, lack of water, improper mineral nutrition or pH and physical injury.

Wilt disease control

The most important control for ornamental vascular wilt diseases is using pathogen-free seeds or cuttings. Using seed certified as pathogen-free will lessen the chance of introducing wilt-causing organisms into a seed bed or field. Since a stock plant may be infected but symptomless at the time cuttings are taken, take propagation stock from "culture-indexed" material (see description in Chapter 11).

Strict sanitation procedures will further reduce the possibility of wilt diseases infecting a crop. When making cuttings, break them out or make them with a blade sterilized by dipping in 70 percent ethyl alcohol. Sterilize the blade between each stock plant. Avoid dipping cuttings in liquid hormone or fungicide suspensions since this could spread wilt pathogens among the cuttings.

Treat soil and potting media with steam or chemicals to eradicate wilt pathogens that might infect the media. Steam treatment is the best method. Apply steam so the coolest portion of the treated batch reaches 180 F for at least 30 minutes. The media should be moist but not wet for best results.

Pasteurizing growing media with mixtures of steam and air has been successful in sanitizing growing media for many years. Mix steam with just enough air to lower its temperature to 140 F. This temperature eradicates soil-borne pathogens but permits beneficial microorganisms to survive. The resulting growing medium supports plant growth and development.

If steam treatment is not possible, use chemicals such as methyl bromide plus chloropicrin. Allow sufficient time for soil to "air out" prior to use. Chemical disinfestation is of questionable effectiveness for eradication of *Verticillium* spp.

Routine sanitation practices in greenhouses and work areas also help control wilts. Destroy old, unhealthy plants. Wash benches, pots and equipment to remove contaminated soil and rinse with a disinfectant. Wash and sanitize cultivation equipment and other tools regularly. Keep all water hose ends off floors.

Follow good horticultural practices to keep crop plants vigorous. Follow proper fertilization schedules to ensure good plant growth. Maintain proper soil or growing media pH. Irrigate to provide an even supply of sufficient water. Avoid splashing water during irrigation to prevent dispersing pathogens among plants.

Managing bacterial vascular wilts with pesticides hasn't been generally effective. Fungal-induced vascular wilts of some container-grown crops can be at least partially managed with systemic fungicide drenches of thiophanate-methyl. Taken into plants, this fungicide acts on the pathogen in the plant's vascular system to suppress symptom development. Recent research with chelated, systemic copper bactericides has shown some promise in managing bacterial wilt pathogens.

Chapter 13

Root and Crown Rots

Root and crown rots are a constant threat to the profitability of flower or foliage plant production for many reasons. Many different fungi can cause root and crown rots. Each pathogen has its own environmental needs. Thus, one or another can appear at any time during a crop's production cycle. Root and crown rots can kill plants in the greenhouse or drastically shorten their usable life after sale. In this way they directly affect the amount of harvested crop. Most importantly, however, root and crown rots debilitate a crop. They can destroy a crop's uniformity and nutritional balance. They can subvert a grower's attempt to schedule a crop for sale on a certain date.

Root rot is a general term used to describe general or localized rotted roots, whether from infectious or noninfectious causes or natural aging. Root rot begins when cortical cells become nonfunctional or die. As cell death continues, the root appears rotten (brown to black discoloration) and the outer cells readily slough off the vascular cylinder (the stele). Root proliferation is greatly reduced because dead root tips generate no new roots. In some cases, the root rotting pathogens continue to invade crown and stem tissues, causing crown rots.

Root rot on young plants with relatively few roots will usually kill the plants. Many of these diseases are called damping-off diseases. When root rot occurs on more mature plants with many roots, the extent of the pathogen's effect on the root system will determine the severity of the symptoms (stunting, discoloration, wilt, death) on the above-ground plant parts. Significant root rot can occur on plants without obvious symptoms on the plant's above-ground portion. When sudden environmental stress occurs on such plants, they may quickly die from the "latent" or hidden infection.

Root rot may be caused by non-infectious factors such as flooding, drought, freezing, excess heat, fertilizer or salts or by other toxic chemicals in the soil. In most, if not all these cases, secondary microorganisms colonize the damaged tissue. These microorganisms can lead to a pathogenic, root rot syndrome.

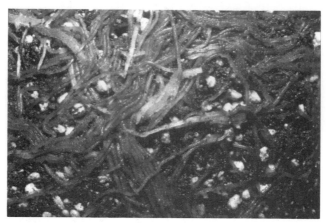

Rotted, brown roots.

Soil-dwelling *Fusarium* spp. and many *Pythium* spp. are commonly found as secondary microorganisms in such cases.

Causes of infectious root and crown rots

Root rot diseases, for the most part, are caused by soil-borne fungal pathogens. Some soil-borne bacteria are causal root rot agents, but are known primarily as causal agents of other diseases.

The fungal genera most frequently involved in root and crown rot diseases of herbaceous ornamentals, generally in order of importance, are Pythium, Rhizoctonia, Phytophthora, Thielaviopsis, Fusarium, Cylindrocladium, Phymatotrichum, Macrophomina, Ramularia, Sclerotium, Sclerotinia and Myrothecium. Many of these fungi are better known for the ultimate disease symptoms they cause (such as damping-off, wilt, crown or stem rot, blight) rather than as root rot pathogens.

Diagnosing infectious root and crown rots

When root pathogens invade and kill the root cortex, the cells become brown, much like leaf or stem tissues brown in response to injury or infection. In some cases, fungal toxins may be responsible for killing and discoloring cortical cells. In other cases, it's simply the pathogen growth into the tissue that kills it. Fungal sporulation may occur in these lesions. Later on, lesions combine to involve major root portions. The root cortex may become so completely rotted that it can readily be slipped off the vascular cylinder which remains as a stiff, thread-like core.

Unlike some foliar pathogens, root pathogens' signs aren't readily visible to the field diagnostician. A microscope is usually necessary to view the resting spores, fruiting structures or mycelial strands in or on the infected root tissue. Sometimes mycelial strands can be noted near the crown of seedlings damped off by Pythium or Rhizoctonia. In most cases, confirming a particular root rotting pathogen has to be done in the laboratory. In the lab, a technician conducts microscopic examination or culturing of the symptomatic tissue.

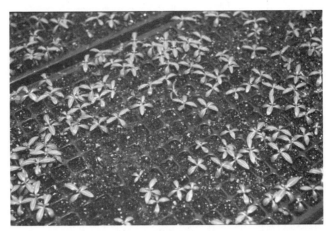

Damping off is root and crown rot of very young plants.

Culturing, which takes seven to 10 days, involves placing a tiny piece of the tissue on a gelatinlike growth media surface. If the suspected pathogen is present, it will usually grow out onto the surface where it can be seen and identified.

Since specific root pathogens are rarely diagnosed in the field, field diagnosis of root rots rests largely on correctly interpreting nonspecific symptoms. A plant exhibits nutrient deficiency symptoms, reduced growth, wilt or death, depending on the number of rotted roots. A plant with 10 percent rotted roots can probably maintain a healthy appearance. If conditions favor continued pathogen development, the plant will soon have 20 to 30 percent or more of its roots rotted, and will begin to exhibit above-ground symptoms. Such plants respond to reduced root capacity by wilting during water stress periods, growing more slowly, developing foliar discoloration typical of nutrient deficiencies, defoliating or dying. If the infection has progressed to the plant crown, plant death usually quickly follows.

When root rot fungi injure or destroy roots, they alter total root function, not just that relating to water or nutrient uptake. Root infection impacts plant growth and development. For example, cytokinins are produced in the root apex and move through the xylem to the shoot. Cytokinins promote cell division, leaf expansion, stolon or shoot initiation, translocation of assimilates and inorganic phosphorus, transpiration and inhibit leaf senescence (lower leaf yellowing). Other growth substances produced in root apices and translocated in the xylem are gibberellins, which promote stem elongation, flowering in some long-day plants, leaf expansion, abscission and fruit growth.

Infectious root and crown rot disease cycles

Because of diverse plant species and cultural situations, as well as diverse pathogens that may be involved, a single set of conditions that consistently affects all root rot disease is difficult to describe. However, there are some common denominators regarding organism sources in production systems,

Root tissue with an internal pathogen.

their functions in parasitic and/or saprophytic phases and the ways most survive to parasitize again.

Survival and spread

All fungal root rot pathogens can exist in the growing medium either as mycelium or spores in or on the infected host tissue, as resting structures (chlamydospores, oospores, resistant hyphal fragments or sclerotia) in the host debris or freely growing in the soil. Most of these organisms are confined to the soil and thus are able to parasitize only roots and/or the lower plant crown. Some, however, can colonize upper plant portions. In such cases, mycelium or spores may grow or be blown, splashed or otherwise carried to other areas or plants. Rhizoctonia web blight, for example, describes a Rhizoctonia infection that has progressed to the plant's upper portions.

Most root rot pathogens exist in the soil or growth medium or on containers or propagative material in an inactive state. When they come into contact with a susceptible host plant, they become active (germinate) and infect the root. Eliminating the inoculum source or preventing this inoculum's activity is the key to controlling most root rot diseases.

Most root diseases result from plants growing in infested soil. The fungus is already in position to contact host roots. The infested soil may have been used directly, mixed with some other noninfested growing media components or come in contact with the growing media surface via some transfer mechanism.

Infested soil may be dispersed by workers on tools, shoes or containers, by wind (as dust), by moving water (rain or irrigation) or by animals. Tiny microscopic soil particles containing propagules of root rot pathogens may initiate root infections. It's difficult to prevent introduction of small, infested soil particles into uncontaminated growing areas. Soil can be dispersed by a watering hose contaminated from being dropped onto infested soil on walkways or by using containers that previously contained infected plants without first removing infested soil residue and thoroughly sanitizing the container.

A vinca crop with black root rot.

Water may be an effective root rot pathogen carrier. Contaminated growing media in water can be splashed by rain or irrigation onto nearby plants or containers. Standing water on a bench or plant bed can harbor root rotting pathogens. In many production areas, irrigation water comes from storage ponds that have collected runoff water from irrigation and precipitation. Runoff water can carry root rot pathogens. Storage ponds may harbor these fungi, especially in areas where winters are mild.

The water mold pathogens Pythium and Phytophthora produce spores (zoospores) that can swim short distances. These organisms are especially troublesome because they may be carried so readily by surface water or capillary water films on mats or in growing media. Water mold zoospores from drain holes of a container holding an infected plant can travel or be washed some distance on bed or bench surfaces and drawn up by capillarity into another container. The plant material itself can harbor root rot pathogens and provide a means of dissemination. Seed transmission of root pathogens is possible, but rare. Flowers or seed pods may touch the soil or become contaminated by dust or soil particles during harvesting. Since many herbaceous ornamentals are propagated vegetatively, cuttings taken from stock plants may carry root pathogens as surface contaminants in growing media splashed up from below. This is especially true of cuttings taken from lower portions of the plants. Occasionally, cuttings may touch the soil or may be placed into contaminated containers for transport to the propagation beds. If they become contaminated, even though not infected, the pathogen can wash onto new roots as they form in the rooting medium.

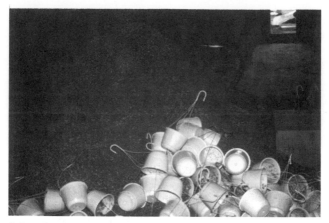

Infested growing media on these old baskets will contaminate
new growing media with root rotting fungi.

When pathogens cause stem or foliar infections as well as root infections, cuttings taken from infected stock plants could be infected with the leaf or stem phase of the disease. When the infected leaf abscises, it becomes a substrate for conidia sporulation that can wash down into the medium and infect newly formed roots. An example is Cylindrocladium of azaleas, which can cause a root rot as well as a leaf spot.

Bulbs or corms may also carry root rot pathogens, especially if they were field-grown in infested soil. Lily bulbs, for example, are harvested from fields by machines that sever the roots below the bulb. Such injured roots may have been or may become infected with Pythium, Fusarium or Rhizoctonia, and that infection may carry over with the bulb as it is potted up for greenhouse forcing. Some of those organisms can also colonize surfaces of bulb scales and be in position to initiate infections on roots as they form.

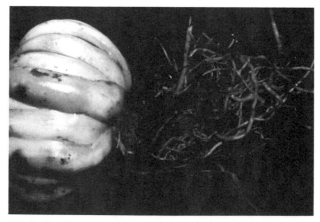

Lily bulbs commonly have root pathogens on them
from field production.

Irrigation water treatment is sometimes necessary to eliminate root pathogens.

Infection

When susceptible host roots are present, pathogen propagules (either resting structures or spores) are stimulated to germinate. The germination process is a response to host root exudates that provide the pathogen food.

Root rot pathogens invade cortical root cells and cause total dysfunction and death of each cell they enter. The cortical cell invasion often progresses from the root surface via runner hyphae that grow along the outer root edges. The pathogens usually produce spores in or on the infected tissue shortly after infection begins. Root surface spores may be carried readily by moving water to other uninfected roots. This process may take hours, days or weeks, depending on environmental conditions, the pathogen and the host's relative susceptibility to disease.

Environmental factors that promote root diseases

Root rot diseases may often remain restricted, more or less, to the original infection site. As a result, a low proportion of the root system may be involved and no visible top symptoms appear. The host may generate new roots at a rate comparable to the pathogen's root-killing rate. If new roots can sustain the plant, no top symptoms are expressed. If environmental conditions generally favor the host, the plants may survive for indefinite periods. An environmental change, possibly as a result of sale or transplant, may favor pathogen development or increase host susceptibility. The result can be badly diseased plants in a short time.

Sanitizing used plant trays.

Stress predisposition to root diseases is an extremely difficult condition to document precisely. It's common to encounter situations where disease suddenly appears after some change in cultural practice or after plants are shipped to another geographical location. Environmental factors most frequently involved in increased root disease include changes in light levels, fertilizer levels, watering practices, soil pH, increased salt levels, pesticide or growth regulator application, repotting to a different sized container, air pollution and insect infestation.

Each root rot pathogen has different temperature, pH and moisture optima and ranges for disease development. Some trends useful as generalizations are: Low soil temperature usually reduces Rhizoctonia root rot, low soil moisture usually inhibits Pythium and Phytophthora and maintaining soil pH below 5.5 generally reduces Thielaviopsis and Pythium root rots. Such information is very useful for root rot control, because disease-suppressing conditions are usually consistent with those needed for good plant growth. This is, again, an example of holistic plant health management. Generally speaking, healthy plants are less susceptible to root rot infection than plants in a state of low vigor from various stresses. The table at the end of this chapter provides some specific conditions relating to many ornamental root rots.

Control

Host susceptibility is the sum of its vigor and genetic characteristics. Many food and fiber crops have been developed with genetic resistance to diseases, including root rots; however, this generally isn't true for ornamental plants. Thus, holistic health management principles take on even more importance.

The best approach to controlling root and crown rot diseases is prevention. One must constantly be aware of a pathogen's potential to get into soil or potting mixes, to contaminate planting stock and containers or to be carried in by water. Although it's difficult to devote time and resources to prevent root rot diseases, it's a good investment. Special cultural practices, if followed routinely, can eliminate root rot pathogens from the production system. If those

These lilies were drenched with fungicide at planting.

practices fail, then you must face the problem of dealing with an established pathogen.

Many, if not most, natural soils contain fungal pathogens capable of causing root rot. Because soils may contain root pathogens, most growers have shifted to soilless potting mixes. Growers sometimes assume that the components in soilless mixes have been treated so they are pathogen-free. This would be true for components produced by a heat process, but it isn't true for other natural ingredients such as peat or bark. Thus, it may be advisable to sanitize even soilless mixes before using them.

Treat soil or potting mixes to control root rot pathogens by heat, chemical fumigation or drench. Method choice will depend on economics, availability of steam and the nature and location of the soil or mix to be treated. Outside ground beds generally can't be readily steam-treated, but can be treated chemically. Media sanitation is discussed in more detail in Chapter 5.

With any soil or potting mix treatment method, if root pathogens such as Pythium and Rhizoctonia aren't completely eliminated by the treatment, they may flourish in the relative biological void created and cause worse disease than if treatment hadn't been applied. Another hazard of the complete or over-kill approach with either heat or chemicals is the ease with which introduced pathogens can become established. Your goal is to eliminate the root pathogens but leave some organisms with potential buffering or antagonism to introduced pathogens. Some composted bark mixes available are specially prepared to retain these beneficial organisms. An antagonistic fungus to add to growing media after sanitation will soon be sold in packets.

Treating irrigation water known or suspected to carry root pathogens is possible and advisable if you're using pond water from field runoff, especially in areas with mild winters. Filter or heat small volumes to remove pathogens; treat larger volumes by injecting chlorine gas into the irrigation line. It's important that the chlorine be in contact with the water for 30 minutes prior to use. Also, determine the needed chlorine amount by calculating the residual chlorine left in the water after 30 minutes. Residual chlorine levels should be around 2.5 parts per million. You can use a swimming pool testing kit for this purpose.

Ebb and flow irrigation systems may harbor root pathogens as well. Research on the proper way to sanitize this water is currently underway. Putting chemicals in the water exposes the plants to chemical contamination at the next watering. Some growers successfully heat water as it returns to the holding area.

Treating used containers is usually necessary to eliminate root pathogens from production areas. An appreciation for infective fungal propagules' small size will make it apparent that containers can't be washed thoroughly enough to remove residual soil or potting mix that may still harbor root pathogens. Many good disinfectants are available for this task. Remember that many chemical disinfectants are readily inactivated by soil, either chemically or physically. Thus, effectively sanitizing dirty containers is difficult. Gas, wet heat or steam treatment ensures that containers are pathogen-free, even if they have a lot of residual growing media on them.

Sanitation principles and practices must extend to other production aspects including tools, hoses, walks and boots. Failure to remember the ways pathogens can invade pathogen-free production systems could cancel out all earlier efforts to have pathogen-free production. Chapter 5 discusses sanitation in more detail.

Excluding root pathogens from production areas also involves planting stock. Seeds, cuttings or bulbs may carry root pathogens if they have been in contact with soil or if they have been contaminated by water-splashed soil or dust containing pathogens. Signs of infection or contamination of such plant material aren't usually apparent. If you know or suspect that propagules carry root pathogens, then use decontamination measures. Use fungicide seed treatments or fungicide dips for cuttings or bulbs. These treatments' effectiveness depends on whether the pathogen is on the surface or in the tissues. In addition, be careful not to inadvertently spread a bacterial pathogen by dipping a plant propagule into a fungicide. The fungicide, of course, would have no effect on the bacteria!

Heat treatment of cuttings or seeds may be useful if the pathogen is less sensitive to heat than the plant propagule. Chemical or heat disinfection may be essential to the initial establishment of pathogen-free propagules. Once stocks are pathogen-free, establish protected propagation blocks for future propagules. Frequently check stocks to guard against recontamination.

Once root pathogens infect production systems to the extent that their exclusion or eradication is not feasible, growers can only hope to suppress the disease incidence and/or severity. Generally, this happens through manipulating cultural practices, applying chemical pesticides or developing or encouraging biological control systems.

Cultural management or control of root pathogens depends on some knowledge of the involved pathogen's biology. The table at the end of this chapter indicates effective practices. For instance, Pythium or Phytophthora are water mold fungi favored by excess soil water. Naturally, avoid those conditions by careful watering and/or by growing the crop in a well-drained medium. Knowing that certain pathogens are encouraged or discouraged by various nutritional, soil pH or temperature conditions allows you to chart a management course to avoid disease. For example, since high ammonium can increase Fusarium root rot incidence, reduce Fusarium disease incidence or severity by limiting ammonium sources of nitrogen.

Chemical fungicides are commonly used to suppress root pathogens. Certain fungicides have a broad disease spectrum, while others are very specific. Some are eradicative, but most act as protectants. If root rot problems are consistently present, a regular, preventive fungicide program may be economically justifiable. Fungicides can be applied as dusts, dips or drenches at the time of propagation and/or at regular intervals thereafter. It's important to know something about the phytotoxicity potential of the chemical being applied as well as the action mode. Most "fungicides" are really "fungistats"; that is, they help to keep the fungus in check, but don't kill its propagules. Don't expect a material to eradicate a soilborne fungus when the medium is drenched. Even with the so-called systemic materials, don't hope to eradicate internal infections that occurred prior to treatment. Chapter 5 discusses the chemical drenches appropriate for root and crown rot disease management.

Cultural practices and fungicides generally suppress or mask root diseases. These practices may effectively reduce the incidence and/or severity of root disease symptoms as long as the suppressive conditions are maintained. If conditions change, suppression may no longer be present and rapid disease development is possible. Thorough disease prevention by eliminating the pathogen greatly reduces the risk of sudden disease development as a crop matures or after it's sold.

Root pathogens in natural soils live in a balanced state with many other organisms, each being limited by the rest. That limitation produces a balance which naturally controls severe root disease. Man's disturbance of the balance, especially in ornamental plant production, reduces the limitation microbes place on root pathogens. Thus, in growing media systems, such as those used to produce flower and foliage crops, there is great need to restore the biological balance to biologically suppress root pathogens.

Establishing beneficial microorganisms in or on herbaceous ornamental roots early in the production cycle is a worthy goal. Evidence suggests tremendous potential for these organisms to help plants grow and repel root pathogens. Future research will provide the technology to introduce inoculum into treated soils and reconstitute natural biological buffers that effectively block root pathogen invasion. New potting mixes containing composted bark are available to suppress root rot pathogens through chemical and biological factors that inhibit normal pathogen behavior. For certain crops, such mixes may have potential for root rot control, however, their effects are not long-lived. Growers must not rely on them solely for root and crown rot protection.

Environmental stress-promoting conditions that predipose ornamentals to some root and crown rot diseases

Pathogen	Stress-promoting condition
Fusarium	Wounds caused by insects or handling. High ammonium.
Pythium irregulare	High soluble salts. Overwatering.
Pythium ultimum	Cool temperatures (<68 F). Overwatering or poor drainage. High soluble salts.
Phytophthora parasitica	High temperatures (>77 F). Overwatering and poor drainage.
Rhizoctonia	Alternating extremes of wetting and drying. High temperatures. Wounds at the plant's crown. High soluble salts.
Thielaviopsis	High pH. Temperature too high or too low for the crop. High ammonium.

Chapter 14

Diseases Caused by Nematodes

Nematodes feed on roots with a needlelike mouth part called a stylet. This feeding usually causes dead areas in root tissues. In addition, root knot nematodes cause abnormal root cell growth. In all cases host plant root functions are greatly decreased, resulting in various, nonspecific symptoms, such as stunting, slow growth, various mineral imbalances and water shortage symptoms. Foliar nematodes invade leaf tissue and cause general blighting.

Plant pathogenic nematodes also interact with other pathogenic microorganisms, causing further disease damage. Pathogenic bacteria may invade plants through nematode feeding injuries. Some viral pathogens have nematode vectors. Nematode interactions with soil borne fungi are common. These fungi invade nematode wounded roots and cause root rotting or vascular wilting.

Many plant parasitic nematodes cause damage on ornamentals (see table below). The most significant damage to herbaceous ornamentals is caused by relatively few genera, however. Meloidogyne, Pratylenchus, Paratylenchus and Aphelenchoides are four of the most important genera on ornamentals.

Plant parasitic nematodes found on ornamentals

Species name	Common name	Species name	Common name
Aphelenchoides	Foliar nematode	Belonolaimus	Sting nematode
Crionemoides	Ring nematode	Ditylenchus	Stem nematode
Heterodera	Cyst nematode	Hoplolaimus	Lance nematode
Longidorus	Needle nematode	Meloidogyne	Root knot nematode
Paratylenchus	Pin nematode	Pratylenchus	Lesion nematode
Radopholus	Burrowing nematode	Rotylenchus	Reniform nematode
Scutellonema	Spiral nematode	Trichodorus	Stubby-root nematode
Tylenchorhynchus	Stunt nematode	Tylenchulus	Citrus nematode
Xiphinema	Dagger nematode		

Aphelenchoides (foliar nematode)

Symptoms

Aphelenchoides can enter leaves through stomata or wounds. They cause a range of symptoms, including brown, dead lesions and overdeveloped red or purple leaf pigments. Some of these symptoms may mimic symptoms caused by sunscald, pesticide damage or high temperature injury.

These nematodes have rapid reproduction rates. Extremely active, they are capable of moving rapidly up stems and across leaves in thin films of water. They are quite resistant to fluctuations in temperature and humidity.

When Aphelenchoides feed in buds, rosetting and growth deformations result. Infected plants' young foliage becomes hardened and stiff to the touch. Some plants, which have a high tolerance to foliar nematodes, can harbor substantial populations with few visible symptoms.

Control

Under moist conditions, bunches of Aphelenchoides form on infected plants' lower leaf surfaces. These nematode masses are easily moved by splashing water or on tools or hands. Careful watering and humidity control are important controls for Aphelenchoides. Removing individual infested leaves is effective for management because most crops aren't infected systemically by Aphelenchoides.

Aphelenchoides survive for extended periods in dried, infested plant debris, particularly when this material isn't subjected to alternating wetting and drying. Plant material adhering to benches, pots and production equipment can serve as inoculum sources for susceptible crops.

Using nematode-free propagating stock is basic to complete control of foliar nematodes. Hot water treatments have been used to eradicate foliar nematodes from foundation stock plants, but this technique is impractical on a large scale. High temperatures frequently disfigure or destroy treated plants. Thus, the practice isn't useful unless plants are grown for stock use or for shoot tips only. Hot water treatments can also be excellent means of disseminating fungal and bacterial plant pathogens, since these pathogens aren't destroyed by the lower temperatures used for nematode eradication.

Systemic nematicides control Aphelenchoides on many ornamentals. Oxamyl applied either as a soil drench, granules mixed with soil, or as a foliar spray is extremely effective in controlling these nematodes on plants for about three weeks.

Using tissue culture techniques to produce clean planting stock can also control Aphelenchoides. If foliar nematodes are accidentally introduced into the callus tissue cultures, they can cause considerable damage. They are readily recognizable in the cultures, however, thus preventing symptomless infestation and dissemination in the callous or plantlets.

Meloidogyne (root knot nematode)

Since most plants are potential hosts for this nematode, root-knot nematodes are often a serious problem in ornamentals. Because the nematodes are within

Foliar nematode damage on chrysanthemum.

host tissue, they are transported easily in rooted cuttings or on symptomless plants. Movement also occurs via dirty machinery, hands, shoes and irrigation water. Once established, these nematodes maintain themselves on a wide range of crop plants and weeds and can become endemic problems.

Symptoms

Infected roots are knotted with small, swollen areas (galls) as a result of the feeding activities of *Meloidogyne* spp. developing within the roots. The type and amount of root knotting depends on the species present, the nematode behavior, the host plant's root characteristics and the nematode population. Warm temperature species of Meloidogyne cause extremely large galls resulting in roots with a tuberous look. Cool weather species usually develop small, discrete galls. Infected plants may react by developing numerous lateral roots above the developing gall. Fine rooted grasses may become infected but may not produce readily visible galls at all. Fleshy, rooted plants may produce extremely large galls, even with low nematode populations.

Above-ground, root-knot nematodes produce the nonspecific symptoms mentioned earlier: signs of nutrient or water deficiency, stunting and slow growth, patchy growth patterns and increased herbicide sensitivity.

Control

The primary means of root-knot nematode control is soil treatment, since the pathogens are soil-borne. Steam disinfestation of soil or potting mix is extremely effective in controlling root-knot nematodes. A wide range of soil fumigants has been available for effectively controlling soil-borne nematodes since the 1940s. Fumigants such as highly refined dichloropropane-dichloropropene mixes, methyl-isothiocyanate, metam-sodium, ethylene dibromide, and various combinations of these materials effectively control Meloidogyne in field soils. Unfortunately, many of these are no longer available for use. Fumigants such as methyl bromide with chloropicrin have been used for many years as effective soil sanitizers for potting mixes and greenhouse or nursery bed soils. Fumigants don't completely sterilize the soil. Low nematode numbers remaining after the treatments can increase to damaging levels in two or

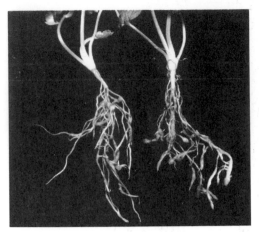

Damage from root knot nematodes.
Photo by Richard M. Riedel, Ohio State University.

three years, especially if nematodes are reintroduced with transplants. Many newer, contact/systemic nematicides are useful as bareroot dips for transplant materials. Hot water dips can also be used for this same purpose.

Pratylenchus (lesion nematode)

Symptoms

Brown and dead flecks or lesions result from primary root feeding by this nematode's larvae or adults. The nonspecific symptoms on above-ground plant parts infested by Pratylenchus are the same as those caused by any factor that damages roots or restricts root development.

Tissue damaged by lesion nematodes is susceptible to soil fungal infection. The lesions not only serve as infection sites but also greatly stress the host and

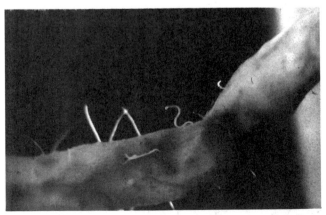

Nematodes feeding on root.
Photo by Richard M. Riedel, Ohio State University.

increase its susceptibility to root rot fungi. These fungi then cause progressive root rot.

Control

Pratylenchus is easily moved by anything that moves soil: workers' shoes and hands, machinery, surface water or wind. The nematodes are moved frequently in host plants themselves. Many ornamentals and weeds can support lesion nematodes. As a result, these nematodes are likely to become a long-term problem once introduced into areas of intensive ornamental production. Using only Pratylenchus-free planting stock is a first step to successfully controlling this nematode. Effective crop rotations to control lesion nematode are difficult since common species have wide host ranges. Weed control is necessary in crop rotation.

Pre-plant soil fumigation is a major control method for lesion nematodes. The nematodes are successfully controlled by the same materials mentioned for root-knot nematode. Post-plant applications of the contact-systemic nematicides control Pratylenchus in plants in containers or closed-bottomed beds. The most severe problem in chemical control is that present product labels for ornamental crops aren't inclusive.

Paratylenchus (pin nematode)

Symptoms

Pin nematodes feed primarily from the outside, either on root hairs or on epidermal cells. Sometimes they penetrate roots and feed on tissues from within. Thus, they produce symptoms similar to those of the lesion nematodes. They have been found by different investigators in many parts of the world, usually in small numbers, and they feed on different kinds of plant roots. Investigators may have underestimated these parasites' abundance and importance. Because of their small size, they may pass through sieves used to separate them from soil in diagnostic laboratories. As many as 10,000 per pound of soil have been found. Many ornamentals' poor growth has been correlated with high numbers of the nematode on bed-grown ornamentals outdoors as well as in the greenhouse.

Control

Control of pin nematode is the same as for the lesion nematode discussed above. In fact, the two nematodes often occur together in one crop.

Chapter 15

Viruses

Viruses have caused increasingly serious problems on ornamental crops in recent years. This is probably due to increased plant movement throughout the world. Like many plant pathogens, viruses can move within an apparently healthy cutting or seed, undetected by plant inspectors.

A virus is an infectious parasite that carries its own genome but needs ribosomes and other components from its host cells for multiplication. This extremely small organism is visible only through the electron microscope.

Plant viruses have traditionally been named on the basis of disease symptoms they cause in the host in which they were originally described. For example, the virus that causes light and dark green areas on dieffenbachia leaves is named dasheen mosaic virus because it was first described on dasheen, a Southern weed.

Many viruses share one or more characteristics often unrevealed by the name. Plant viruses are thus grouped according to their physical and chemical structure. The grouping includes shape and size, immunological properties, natural transmission method and host reactions. Currently, plant viruses are divided into 23 groups. Sixteen of the viral groups are known to include viruses infecting ornamentals. Five of these groups have elongated particles and eight groups have particles with an isometric or almost spherical shape.

Disease cycles

The disease process begins when the virus enters the cell through a tiny wound. These wounds allow viral particle entry but do not kill the cells. Wounds may result from physically scraping the leaf surface with some outside object or from insect feeding. Insects may carry viral particles from plant to plant as they feed on diseased plants and then feed on healthy ones. Once the virus has entered the cell, it begins multiplying. The host may react in several ways.

Viral multiplication results in the virus spreading from the originally infected cell to neighboring cells. In local infections, affected host cells die after very

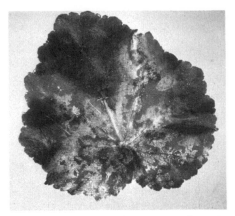

Mosaic and ring spots from virus in begonia.
Photo by Larry W. Barnes, Texas A&M University.

limited cell-to-cell viral spread. A brown, dead spot is all that remains of the viral infection. Usually, with important ornamental viruses, the entire plant eventually becomes invaded and symptomatic. Such a situation is called a systemic infection. Viruses generally survive through systemic infections. As systemic pathogens, they may be picked up by an insect or disseminated by an infected cutting or seed or on workers' hands or tools. It's important to know which virus is in a plant to know exactly how it's able to spread and infect other susceptible plants.

Diagnosing viral diseases

A localized infection may develop at the infection site. It may be a yellow, brown or red spot. The tissue may be dried out as well as discolored. These infections may begin as small spots of affected tissue that gradually grow larger as the virus moves slowly from cell to cell. As discussed above, tissue in the

Viral mottling in gloxinia.

Viral-induced tissue death on gloxinia.
Photo by Larry W. Barnes, Texas A&M University.

lesions or spots may die, preventing further spread of the infection. In other cases, local infections simply precede systemic spread and systemic symptoms.

Most plant viruses with economic importance infect the host and spread from the infection site to the vascular tissues where they move throughout the plant. The first symptom of such a viral infection may be associated with the vascular system. Symptoms associated with the vascular tissue may include vein yellowing or vein "banding." Vein banding is a clearing or yellowing of leaf cells along the veins, giving a widening effect. Vein death may appear in sensitive species, sometimes resulting in death of a stem section or the entire plant.

Symptoms don't always increase in severity in the virus-infected plant. During the initial infection process, severe symptoms develop and the host may die. This is a shock phase or acute phase. If the host survives, a chronic phase ensues in which some of the affected parts show recovery. New growth may appear with less severe symptoms. Eventually, the plant may develop without any symptoms. This is called a latent phase or latent infection.

In any plant, symptom visibility may relate to how long the plant has been infected, the plant's age at time of infection, the viral strain and the infected plant's environment. For instance, viral infection may occur without any visible symptoms, even initially after infection. These latent infections are a problem in controlling viral spread. Many weeds can also be infected by viruses without symptoms.

Viral infection often reduces ornamental plant quality. Stunting is a general symptom of many virus-infected plants. Affected plants may have smaller leaves, reduced internode length and smaller flowers.

Yellowing is a symptom that may be uniform in the virus-infected plant or may be confined to a leaf portion or leaf veins. Yellowing occurs primarily along a leaf edge, usually in newly expanding leaves. It's difficult to know whether this yellowing is due to viral infection or to cultural and environmental situations.

Virus-caused color breaking in tulip.

Mosaic symptoms are produced by patches of normally colored, uninfected cells intermingled with abnormally colored, infected cells. Small veins usually delineate the patches. The mosaic appears as elongated patches or streaks in leaves with parallel veins, such as those on lilies, iris or narcissus.

Mottling is abnormal coloration in the leaf that doesn't follow the veins. Patches aren't sharply bordered. They develop across vein boundaries. Sizes and shapes of discolored areas vary. Often, such a symptom is called speckling, stippling, spotting or blotching.

Viral-induced tissue death can range from mild leaf browning or etching to severe browning of large leaf or stem areas. Browning from viruses can appear as rings, spots or darkening of leaf or stem veins. In very susceptible plants, dead growing points may appear as viruses move systemically toward the top of the plant. An example would be the reaction of gloxinia to tomato spotted wilt virus. Small, systemic dead spots are more common. They appear as brown spots or flecks on leaves, such as the common fleck viral infection found on Easter lilies.

Several viruses infesting ornamentals produce color breaking or color variegation on flowers. Viruses can cause either light streaks, dark streaks or flecks. Some viruses cause a spotting of dark or light color in flowers. Flower greening is a modification of flower color frequently associated with mycoplasmalike organisms. Some viruses also produce flower death or brown spotting.

Diagnosing viral diseases is usually not complete when based on visual symptoms alone. Bioassay, serology, nucleic acid analysis and electron microscopy are procedures used for laboratory virus identification.

Many viruses infecting ornamentals can be mechanically transmitted. Sap from the suspect plant is rubbed onto the leaves of the assay host (a bioassay). The resulting infection's characteristics help identify the virus. Serological tests are usually used to confirm visual diagnosis or results from mechanical transmission tests. Recently, using serology in routine indexing has expanded with the enzyme-linked, immunosorbent assay (ELISA) procedure. Under elec-

Screening used to prevent insects from invading the greenhouse.

tron microscopy, viral particles are visible in sap extracts from infected plants. The virus can be recognized and tentatively identified on the basis of its size and shape. Finally, a new technique based on isolating and identifying the nucleic acids present in symptomatic tissue can help determine which virus is present in the diseased plant.

Plant virus spread

As noted above, viruses must initially penetrate and enter the plant through a wounded cell. Depending on the virus, however, efficient transmission can occur through seed or vegetative propagation of plant parts after the initial infection is established. Seed transmission of plant viruses can introduce into a crop a virus that then serves as an infection source for transmission by mechanical means, insects or nematodes. The percentage of seed infected by a virus may vary from a few hundredths of a percent to 100 percent. Viruses may persist in seed for long periods, thus permitting transport over great distances. The proportion of seed infected depends on several factors. These include the viral strain, host plant, time at which the plant is infected, location of seed on the plant, age of seed and temperature. Viral groups in which transmission is known to occur in or on seed include the tobraviruses, tobamoviruses, potyviruses, nepoviruses, cucumoviruses and ilarviruses. Certain nematodes can pick up viruses as they feed on infected plant roots. Nepoviruses and tobraviruses can be acquired in 15 minutes to one hour and are inoculated in a similar period. Nematodes retain both groups of viruses for many weeks and both larvae and adults transmit them. Virus carrying nematodes aren't common in greenhouse soils or potting mixes, but cause problems in field-grown, woody perennials, especially in the South or West.

Aphids, leafhoppers, mites, thrips, mealybugs and beetles can transmit plant viruses. Aphids are important vectors of viruses of herbaceous ornamentals. Viruses in the carlavirus, potyvirus, closterovirus, cucumovirus and caulimovirus groups are aphid-transmitted.

Viruses that are aphid-transmitted can be separated into groups by the length of time the virus persists in the aphids. Nonpersistent viruses usually cause mosaic symptoms. These viruses can be acquired in as short a time as 10 to 30 seconds. Inoculation occurs just as quickly. With many nonpersistent viruses, aphids become noninfective very rapidly.

Semipersistent, aphid-transmitted viruses persist longer in the vector than non-persistent viruses. The probability of viral transmission with these viruses increases with feeding periods on affected plants up to four hours.

Persistent, or circulative viruses, have minimum aphid acquisition and inoculation feeding periods of 10 to 60 minutes. They often have a latent period of 12 hours or more, during which an aphid that acquires a virus is unable to transmit it. Persistent viruses are retained in the aphid and the insect remains infective after molt. No ornamental viruses have been shown to be transmitted this way.

Western flower thrips (*Frankliniella occidentalis*) and two other species of thrips transmit tomato spotted wilt virus. One strain of this virus has been particularly serious on greenhouse crops in recent years. The virus is widely distributed in the United States and has a wide host range. The viral strain found most commonly in greenhouses is vectored persistently by the thrips. The thrips acquire the virus as a larvae and transmit it throughout their adult life (up to 45 days). They apparently don't pass the virus through their eggs to their offspring.

The sweet potato whitefly (*Bemisia tabaci*) has been known for many years to be a vector of economically important plant viruses in tropical and subtropical areas. Infectious whiteflies have induced some 70 or more diseases by feeding. The sweet potato whitefly's range has been increasing in the last few years into greenhouses throughout the world. They are known to transmit many geminiviruses in a persistent manner, closteroviruses in a semi-persistent way and carlaviruses nonpersistently. These insects also transmit other undescribed viruses. Their full threat as economically important greenhouse crop viral vectors remains to be seen.

Intensive cultivation of greenhouse crops with plants closely spaced and often touching each other favors mechanical viral transmission. Of even greater concern, however, is the viral spread on cutting tools, hands, pots and potting surfaces. Without disinfesting knives used to make cuttings, harvest flowers or divide plants, many viruses spread rapidly from plant to plant.

As mentioned above, a virus residing in a weed is often not obvious because most weed infections are latent. Thus, a virus present in weeds may become a disease source when a susceptible ornamental crop is introduced nearby. Also, insect vectors move freely from weeds to crop. A similar situation also may occur when a virus not causing severe economic damage in one greenhouse or nursery plant is transmitted to a newly introduced crop with greater sensitivity to the virus.

The relative efficiency of virus spread depends on the biology of the vectors. For example, the numbers of thrips entering or moving in a crop may vary from year to year and month to month. Weather conditions in winter will have a major influence on thrips survival and multiplication. Nematode vectors in soil generally move slowly and the viruses they transmit spread for only limited distances.

Control of viral diseases

Viral diseases are managed by prevention. This involves growing crops from virus-free seeds or stocks and stopping the viruses from entering the crop or from spreading once infection is visible. Reduce viral spread by using resistant varieties or by reducing vector populations with insecticides.

In today's ornamentals industry, intensive efforts to produce propagation plants free of known pathogens, fungi, bacteria and viruses are underway. This approach to clean stock production depends on rapid and specific detection methods. It also depends on developing methods to eliminate viruses known to be in proposed stock plants. Using these approaches permits establishment of virus-free plants usable as stock plants. The very damaging chrysanthemum and carnation viruses have been removed from commercial varieties by such procedures and are rarely seen today.

Virus indexing is the testing process used to determine if a virus is present in a plant and to identify it. The choice of indexing method depends on the crop and the growth stage when a test is performed. Since indexing has a high unit cost, tests are usually applied to a limited number of nuclear stock plants, as described under culture indexing in Chapter 11.

If no virus-free plant material is available to start a clean stock program, several techniques can eliminate viruses from plants. These are based on exploiting erratic virus distribution within the plant.

Propagation of small tissue pieces, usually shoot tips, from some vegetative branches may yield virus-free plants. Heat treatment or thermotherapy has been successfully combined with removing small tip cuttings or shoot-tip culture to obtain healthy plants. The most common method of producing virus-free explants is to heat treat the virus-infected source plants before excising tissue. The infected plant is exposed to 95 to 104 F for varying time periods, depending on the virus and plant species or cultivar. Treatment normally involves exposing actively growing plants to hot air for three to 12 weeks and occasionally longer.

One of the most effective ways to limit viral spread and thus "control" the disease is to eliminate the insect vector from the cropping area. Installing insect-proof screens over fan intake vents or passive ventilation devices keeps insects away from susceptible crops. Install such screens carefully to prevent tiny insects like thrips from getting in, while still permitting sufficient air flow to cool the greenhouse. In spite of difficulties, the screens are a cost-effective alternative to the intensive, insecticide-spraying programs needed once a vector like thrips is found on the susceptible crop. When a virus that is spread persistently by an insect is detected in a crop, the vector population must be reduced to extremely low levels if not eliminated entirely.

Some plant viruses infecting
ornamental plants

Group	Virus name	Some host plants
Carlavirus (rod shaped, aphid transmitted)	Cactus virus 2	*Cactus* spp.
	Carnation latent	*Dianthus caryophyllus*
	Chrysanthemum B	Chrysanthemum
	Helenium virus S	Helenium
	Hippeastrum latent	*Hippeastrum hybridum*
	Honeysuckle latent	*Lonicera japonica, L. periclymenum*
	Lily symptomless	Lilium
	Narcissus latent	Narcissus
	Nasturtium mosaic	Nasturtium
	Nerine latent	*Nerine bowdenii, N. sarniensis*
	Passiflora latent	*Passiflora caerulea, P. suberosa*
Caulimovirus (spherical DNA containing, aphid transmitted)	Alfalfa mosaic	*Daphne* spp., *Lupinus* spp., *Viburnum opulus*
	Carnation etched ring	*Dianthus caryophyllus*
	Dahlia mosaic	*Dahlia* spp.
	Tobacco necrosis	*Primula* spp., *Tulipa* spp.
Closterovirus (long rods, aphid transmitted)	Carnation necrotic fleck	*Dianthus caryophyllus*
	Carnation yellow fleck	*Dianthus caryophyllus*
Cucumovirus (spherical, seed and aphid transmitted)	Chrysanthemum aspermy (tomato aspermy)	*Chrysanthemum morifolium*
	Cucumber mosaic	*Begonia tuberhybrida, Canna generalis, Celisia argentea, Dahlia* spp., *Dieffenbachia picta, Gladiolus* spp., *Hydrangea macrophylla, Iris* spp., *Lilium* spp., *Lupinus augustifolius, Narcissus tazetta, Pelargonium zonale, Peperomia magnolifolis, Salvia splendens,* many others

Geminivirus (spheres, in joined pairs, DNA containing, transmitted by leafhoppers and whiteflies)	Abutilon mosaic Ageratum yellow vein Balsam leaf curl Euphorbia mosaic Hibiscus leaf curl Hollyhock yellow mosaic Zinnia leaf curl	Abutilon Ageratum Impatiens Poinsettia Hibiscus Hollyhock Zinnia
Ilarvirus (spherical with several component particles, mechanical and pollen transmission)	Apple mosaic Prunus necrotic ring- spot Tobacco streak	*Rosa* spp. *Rosa* spp. *Dahlia* spp., *Rosa* spp.
Nepovirus (spherical, transmitted mechanically, by seed and by aphids)	Arabis mosaic Raspberry ringspot Tobacco ringspot Tomato ringspot	*Narcissus* spp. *Narcissus* spp. *Pelargonium hortorum* *Hydrangea macrophylla*, *Pelargonium hortorum*
Potexvirus (rod shaped, mechanically transmitted)	Cactus X Cymbidium mosaic Hydrangea ringspot Nandina mosaic Narcissus mosaic Nerine virus X Viola mottle Zygocactus virus X	*Cactus* spp. Cattleya, cymbidium *Hydrangea macrophylla* *Nandina domestica* *Narcissus* *Agapanthus praecox* subsp. *orientalis*, *Nerine manselli*, *N. sariensis* *Viola adorata* *Zygocactus*

Potyvirus (flexuous rods, aphid and seed transmitted)	Bean yellow mosaic
	
	Bearded iris mosaic
	Carnation vein
	  mottle
	Dasheen mosaic
	
	Freesia mosaic
	Helenium virus Y
	Hippeastrum
	  mosaic
	Iris mild mosaic
	Narcissus yellow
	  stripe
	Statice virus Y
	Tulip breaking
	Turnip mosaic
		Freesia, gladiolus, *Iris hollandica*, *Lupinus luteus*
		Iris susiana
		Dianthus caryophyllus
		Aglaonema spp., *Caladium hortulanum*, *Colocasia* spp., *Dieffenbachia* spp., *Xanthosoma* spp., *Zantedeschia* spp.
		Freesia spp.
		Helenium
		Hippeastrum equestre, *H. hybridum*
		Iris hollandica, *Iris xiphioides*
		Narcissus spp., *Nerine bowdenii*
		Limonium sinuatum
		Lilium, *Tulipa* spp.
		Anemone coronaria, *Limonium perezii*, *Matthiola bicornis*, *Petunia hybrida*, *Tropaeloum officinale*, *Zinnia elegans*

Rhabdovirus (bullet-shaped, with lipid envelope, phalaenopsis, transmitted by mites, leafhoppers and aphids)	(Several ornamentals contain rhabdovirus-like particles; they have not been named) Orchid fleck	*Chrysanthemum* spp., *Dendrobium hybrid, Dianthus caryophyllus, Gerbera* spp., *Iris germanica* *Cymbidium, dendrobium, odontoglossum*
Tobamovirus (short rods, mechanically transmitted)	Tobacco mosaic Odontoglossum ringspot (orchid tobacco mosaic)	Gerbera, hippeastrum, petunia Cattleya, cymbidium, Phalaenopsis, vanda
Tobravirus (short rods, mechanically transmitted)	Tobacco rattle	Aster, crocus, gladiolus, hyacinth, lilium, narcissus, tulip
Tomato spotted wilt virus (spherical with lipid envelope, transmitted by thrips and seed)	(No other known member, but several strains)	*Ageratum houstonianum, Begonia* spp., *Calendula officinalis, Cyclamen* spp., *Dahlia variabilis, Gerbera* spp., *Gladiolus* spp., *Impatiens* spp., *Lupinus* spp., *Zinnia elegans,* others
Tombusvirus (spherical, transmitted mechanically and "through soil")	Carnation mottle Carnation ringspot Cymbidium ringspot Pelargonium leaf curl Petunia asteroid mosaic	*Dianthus caryophyllus Dianthus caryophyllus Pelargonium zonale, Tolmiea mengiessi, Tulipa fosteriana Pelargonium peltatum* *Petunia* spp.
Tymovirus (spherical, transmitted mechanically and by beetles)	Poinsettia mosaic Scrophularia mottle	*Euphorbia pulcherrima Scrophularia modosa*
Viroids	Chrysanthemum stunt Chrysanthemum chlorotic mottle	*Chrysanthemum morifolium* *Chrysanthemum morifolium*

Chapter 16

Major Insect and Mite Pests

Floricultural crop production for export and domestic consumption is a major component of agricultural income in several countries. Because similar crops are being produced, a fairly consistent group of insects, mites and related pests attacks these crops. The insect pests represent many orders, including Homoptera (aphids, whiteflies, mealybugs and scales), Hemiptera (plant bugs), Diptera (leafminers, midges, shore flies and fungus gnats), Thysanoptera (thrips), Lepidoptera (caterpillars) and Coleoptera (beetles, grubs). Noninsect pests include Collembola (springtails), Acari (mites), Diplopoda (millipedes), Symphyla (symphylans), Isopoda (sowbugs) and Mollusca (slugs, snails).

Occasionally, certain pests within the above group become severe problems and may limit production of one or more crops. Some recent examples include the serpentine leafminers, *(Liriomyza trifolii* and *L. huidobrensis*), western flower thrips *(Frankliniella occidentalis)*, sweet potato whitefly *(Bemisia tabaci)* and beet armyworm *(Spodoptera exigua)*. These pests, along with continued problems with other insect and mite species including aphids, particularly green peach aphids *(Myzus persicae)*, greenhouse whiteflies *(Trialeurodes vaporariorum)* and two-spotted spider mites *(Tetranychus urticae)*, have provided serious challenges to growers, researchers and pest control advisors as they attempt to develop management programs. None of these pests is particularly new. Extension and research bulletins written between 1900 and 1930 mention problems with most pest groups that cause problems today.

The growing severity of pest problems and their resulting spread to areas far from their normal ranges is due to one or more of the following factors:

1. Inadequate biological knowledge.
2. Worldwide plant production and distribution system.
3. Pesticide resistance.
4. Failure of quarantine.
5. Improper pesticide management.

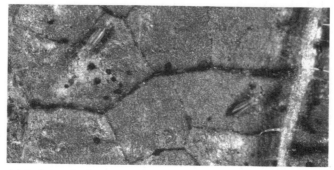

Adult thrips, feeding injury and dark fecal spots on leaf underside.

All factors are interrelated and can contribute to an insect or mite becoming a major problem, but pesticide resistance, improper pesticide management and plant distribution probably cause the most immediate problems. The literature is full of reported pest outbreaks following pesticide overuse. Resistance to one or more pesticides has been documented in at least one species in each major pest group to be discussed here. Many floricultural crops are produced by specialty propagators that carry on intensive pesticide programs to ensure that pest-free plant material is shipped to producers. This intensive program can create pesticide-resistant pests. If plants harboring pesticide-resistant pests are selected at the propagation site, these pests will soon appear wherever the plants are shipped. Selection for resistance continues at the destination.

The following section details the general biology, economic importance and management of many major and minor insect, mite and related floricultural crop pests. Major pests are covered in more detail, but coverage is in no way complete for any pest. The biology of individual species within each group often varies significantly, but the general aspects of biology related to pest management are emphasized for each pest group. The discussion and accompanying photographs should be sufficient to diagnose many problems, but you

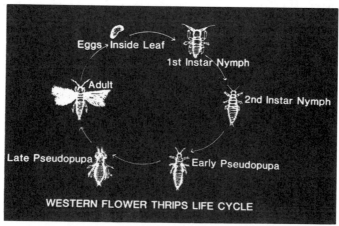

Thrips life cycle. Pseudopupal stages often occur off the plants.

Immature thrips and feeding injury on chrysanthemum leaf.

should contact your extension entomologist for specific identification, as well as current information on the biology and management of these or other pests.

Thrips (Thysanoptera - Thripidae)

There may be as many as 5,000 thrips species, with several hundred species attacking cultivated plants. Only a few species cause problems on floricultural crops, but those problems can be severe. Major floricultural crop pests include the onion or tobacco thrips *(Thrips tabaci)*, greenhouse thrips *(Heliothrips haemorrhoidalis)*, eastern flower thrips *(Frankliniella tritici)*, and the western flower thrips *(F. occidentalis)*. The western flower thrips is an excellent example of a formerly secondary pest that has become extremely difficult to manage and has spread outside its native area due to previously mentioned factors.

Thrips feeding injury on salvia flowers.

Predatory mite (Amblyseius spp.) *feeding on immature thrips.*
Photo by Marilyn Steiner, Alberta Environmental Centre.

Another thrips species causing problems on numerous crops in Japan, Hawaii and several Caribbean islands is the melon thrips (*Thrips palmi*). This thrips has recently been detected in Florida.

Biology

Thrips are very small, narrow insects, less than 1 millimeter wide and only a few millimeters long. Adults have long, fringed wings but aren't strong fliers. Their small size, however, allows them to move long distances on wind currents. Females predominate in most species, and males have never been found for some species. Most of this discussion is based on the western flower thrips.

The western flower thrips female places her eggs within plant tissue. This may be leaf or flower tissue. Depending on temperatures, eggs hatch in three to four days into small, translucent larvae. After one or two days the first-instar larvae molt into second-instar larvae, which generally become yellow as they feed. These first larval instars are usually found protected within developing leaves or flowers. After two to four days the second-instar larvae often move off the plant to the soil or growing medium and undergo two transformation stages, called pseudopupae. The first transformation stage is called the prepupa, and the second is the pupa. These stages can also occur in flowers, so thrips don't always leave plants. Thrips don't feed or move much during this time, which can range from two to five days. Adults then emerge. At normal greenhouse temperatures, the egg-to-adult cycle can be completed in eight to 13 days. Thrips do best in hot, dry conditions. High humidity doesn't seem to reduce their numbers, but relatively constant moisture on leaves and flowers does. Female western flower thrips can live from 30 to 45 days and lay 150 to 300 eggs. Females don't have to mate in order to produce eggs, but eggs produced in this way hatch into males.

The rapid life cycle, high reproductive capacity and the fact that only the larvae, concealed within plant parts much of the time, are very susceptible to pesticides, lead to great difficulties in detection and management.

Economic impact
The western flower thrips has a very wide host plant range that includes many ornamental crops. Thrips damage plants directly by feeding. They aren't true piercing-sucking insects, such as aphids and whiteflies, but feed by inserting their modified left mandible into plant tissue and removing plant fluids with a needle-like stylet. Much of the injury caused by larval feeding occurs within developing leaves and flowers and isn't seen until affected parts expand. These areas are often distorted, and affected flowers may not open normally, but because the injury occurred earlier, thrips may not be present. Thrips also feed on upper and lower surfaces of expanded leaves and flowers. This feeding damage appears as silvery areas on leaves.

Feeding on flowers results in petal scarring. Depending upon petal color, the injury may result in light or dark areas. Usually, feeding injury is accompanied by numerous dark spots of fecal material. Thrips also feed on pollen and nectar of plants such as African violets, causing premature senescence.

A very serious additional problem with some thrips species is the transmission of the virus causing spotted wilt (TSWV—see discussion in Chapter 15). Several thrips species are vectors, including onion thrips, western flower thrips and possibly melon thrips. The virus occurs almost everywhere, and plants in more than 30 families of ornamental and vegetable crops can act as virus reservoirs. Thrips acquire the virus as larvae, in as little as 30 minutes of feeding on infected plants and can become infective after three to 10 days. Usually, adult thrips transmit the virus and remain infective throughout their lives. Apparently there are geographical differences in the ability of the same thrips species to transmit TSWV. No cure for TSWV exists, so managing this disease depends on managing thrips and/or changing crop production practices.

Detection
Monitoring with sticky traps (blue or white are best, yellow is acceptable) is necessary for a successful control program. Thrips are usually seen on traps before becoming visible on plants. Thrips injury is often not visible until leaves and flower parts mature. Thus, it's very important to detect infestations as soon as possible. In greenhouses, place traps vertically just above crops known to be attractive to thrips and/or very susceptible to TSWV. The number of traps to use will vary, but one trap for every 10,000 square feet in a large greenhouse should provide a reasonable estimate of population trends. In a smaller area (a 30- by 100-foot greenhouse) use three traps, one at each end and one in the center. Check the traps at least weekly. Learn to recognize these very small insects as they appear on traps. Keep records of numbers caught and begin to relate these counts to numbers seen on plants. Inspect plants in all areas weekly. Look for thrips injury on leaves or flowers. Shake some flowers over white paper to see if thrips are present. Blowing gently into a flower will usually activate thrips.

Cultural/physical control

Because of their flight habits and wide host plant range, it's difficult to remove enough vegetation from around the production area to lower thrips populations, but doing so certainly won't be harmful and contributes to an overall, pest management program. No flowering plants should be permitted to grow in areas immediately adjacent to the production area. Attempts to use screens or barriers to protect valuable stock plants and/or plants especially prone to TSWV have been quite successful in restricting thrips movement. Because of the fine mesh size required to prevent thrips movement, however, air circulation can be restricted, causing heating and moisture problems. Increased surface area and/or fan capacity will be necessary to counteract these problems unless the barriers are removed in warm weather.

Crop rotation may be necessary to disrupt thrips populations. Changing to less-susceptible crops, immediately harvesting crop residues or harvesting an entire greenhouse area at one time will help reduce thrips numbers. One of the worst situations is a crop production system in a large greenhouse that always contains plants of different ages. Thrips simply move from one area to another. Because of the rapid life cycle, crop residues remaining in the greenhouse or field are excellent reservoirs for increasing thrips populations. Crops extremely susceptible to thrips and/or spotted wilt virus should be produced in separate greenhouses or fields.

Pesticides

In general, control with insecticides is difficult. Many thrips species can survive on a wide range of host plants, both cultivated and wild, which can provide sources of infestation and reinfestation. The most susceptible larval stages are only present for about one-third of the egg-to-adult cycle. Contacting thrips deep inside plant growing points or flower buds is very difficult, so current control strategies are based on frequent (three to five day intervals) applications, using high and low volume sprays, aerosols and fogs, to protect developing flowers and foliage. Using high volume sprays as late in the day as possible keeps plants wet longer and increases the chance of the spray deposit contacting the thrips.

Combining a pyrethroid insecticide with an organophosphate or carbamate insecticide has been successful in some management programs. The pyrethroid insecticide causes the thrips to become more active, increasing the chance of sufficient pesticide contact to cause death. Using sugar with an insecticide also may increase control. The sugar acts as a "bait." When using the sugar plus insecticide, use only about 25 percent of the normal spray volume, not a wet spray.

Possible ways to eliminate thrips (and other pests) include post-harvest fumigation of flowers with methyl bromide, gamma radiation and modified atmospheres (low oxygen or high carbon dioxide). Although none of these methods has been completely successful, they can be used in some situations.

Biological control

Most research involves introducing predatory phytoseiid mites *Amblyseius cucumeris* and *A. barkeri*. One or both of these mite species are being reared and distributed by commercial insectaries and research institutions. Successful control of onion thrips has been reported on greenhouse peppers in the Netherlands. Results against the western flower thrips are varied. Much more work on release rates and intervals is necessary before final evaluation. If compatible pesticides are used or the predators are pesticide-resistant, predatory mites could be useful in an integrated management program.

Predatory anthocorids, such as *Orius tristicolor* and other *Orius* species (minute pirate bugs), have been very effective thrips predators, including the onion thrips and western flower thrips. These predators will move into areas (including greenhouses) if no pesticides are being used, but only after thrips populations have reached high levels. No reports are available concerning the rearing and programmed release of these insects, but some insectaries and researchers are developing rearing and introduction programs.

Some fungi attack thrips. *Entomophthora thripidum* has attacked thrips in several areas of the world. *Verticillium lecanii* has been evaluated in commercial and experimental formulations. To date, no commercial fungal preparations are available for thrips control.

Leafminers (Diptera - Agromyzidae)

Of the approximately 150 agromyzid species known to feed on cultivated plants, only about a dozen species have a wide enough host plant range to cause problems on many protected and outdoor crops. Some of the best known species include *Liriomyza sativae*, *L. trifolii*, *L. bryoniae*, *L. huidobrensis* and *Chromatomyia syngenesiae*. In the late 1970s *L. trifolii* caused serious problems in North American greenhouses, mainly on chrysanthemums, gypsophila, gerbera and bedding plants. Shortly afterward, infestations were reported on

Leafminer (Liriomyza trifolii) *adult.*

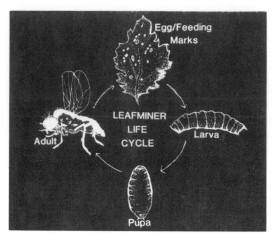

*Leafminer life cycle. In many species, pupal
stage occurs off the plant.*

these same crops in other areas of the world. *L. trifolii* apparently spread on
infested plant material.

In 1989 and 1990 *L. huidobrensis* was found infesting chrysanthemum,
petunia and many vegetable crops in Central and South America. Infestations
of this insect have also occurred in several northern European countries. The
spread to Europe probably was on infested plant material from the Americas.
To date, no serious problems with this insect have occurred in the United States
or Canada.

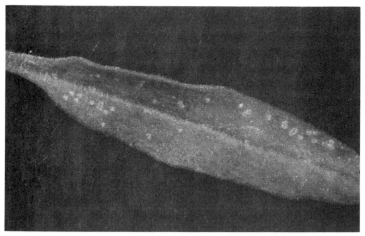

*Feeding and egg-laying punctures made by adult female leafminers in
cotyledon leaf.*

Leaf mine and larva of Liriomyza trifolii.

Biology

Although the specifics of biology vary with each species, general aspects related to management are similar. Leafminer adults are small, black and yellow flies, about the size of fruit flies. Females make small feeding and egg-laying punctures in leaves with the ovipositor. Both males and females feed on fluids that ooze from the leaves. Each female can produce several hundred eggs, depending on the host plant and its nutritional condition. The eggs hatch into small larvae in four or five days. The larvae feed within the leaves, forming characteristic narrow trails or mines. Larvae become yellow when about half grown. The leaf mines of *L. trifolii* are easily seen on upper leaf surfaces, but the mines of *L. huidobrensis* are sometimes on the lower surface (for example, on chrysanthemum) and aren't readily visible from above. Larvae pass through three larval instars in four to seven days, then emerge from the leaves and drop off plants to pupate. The pupal stage lasts about nine days, after which adults emerge. As with other insect and mite pests, the length of the developmental cycle depends greatly on temperature and the host plant.

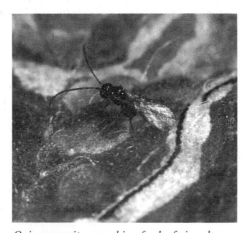

Opius parasite searching for leafminer larva.

Economic impact

On floricultural crops, quarantine regulations instituted by several countries to prevent establishment of *L. trifolii* (or other species) make any injury very important economically. Other economic injury is difficult to measure unless the crop is completely destroyed. In California, however, *L. trifolii* was estimated to have caused a 23 to 27 percent loss to the chrysanthemum crop from 1982 to 1986. In addition to direct injury, growers spent more than $15 million on control programs. Although a very serious problem in some areas, the impact of *L. huidobrensis* has not yet been assessed.

Detection

Inspect plants weekly for leaf punctures or leaf mines. Pay special attention to new plant shipments (cuttings). Monitor adult populations with yellow sticky traps to detect infestations and observe the success or failure of management schemes. For monitoring purposes, space traps approximately 45 to 50 feet apart. Place traps at or slightly above crop height and concentrate them near susceptible cultivars and main entry points into the crop production area. It may be possible, using a sufficiently high trap density, to reduce leafminer populations in a confined greenhouse area, but this isn't generally practical.

Cultural/physical control

Screens or other barriers can prevent leafminer invasion or movement among crops. Some growers have used vacuum suction devices to remove leafminer adults and reduce future larval injury. These devices capture large numbers of adults, but little data is available to indicate the methods' effectiveness.

Treating rooted chrysanthemum cuttings with a combination of cold storage (34 to 36 F) and methyl bromide fumigation has been effective against *L. trifolii* eggs and larvae but less effective against pupae. Plant nutrition is also important. More leafminers may occur on plants receiving higher amounts of nitrogen.

Relatively simple things such as removing weed host plants near the producing crop will help a great deal. Grouping cultivars and/or host plants known to be susceptible to leafminers will aid in chemical or biological control.

Pesticides

Leafminers reach severe pest status only if there is pesticide resistance or poor pesticide management. Many, if not most, leafminer outbreaks are the direct result of using insecticides that destroy natural controls and select resistant populations. The insecticide methomyl (Lannate) has been associated with leafminer outbreaks in California. These insects apparently have a great ability to develop pesticide resistance. In Florida, most insecticides used for *L. trifolii* control had a useful life of approximately two to three years before they became ineffective. The rapid appearance of pyrethroid resistance among leafminer populations may have been due to widespread use of DDT and other insecticides in previous years. This resistance has been characterized as the "kdr," or "knockdown resistant" type, which involves the nerve membranes and is very stable.

Recently, pesticides that have action modes different from conventional pesticides (abamectin, cyromazine, Margosan-O) have become available and provide excellent control. These materials must be applied frequently to be successful and frequent applications may eventually lead to resistance development. In fact, leafminers in Florida may have already developed resistance to cyromazine.

Biological control

Most experimental, biological control programs for leafminers involve one or more hymenopterous parasites. The genera *Dacnusa, Opius* (Braconidae), *Diglyphus, Chrysocharis, Chrysonotomyia* (Eulophidae) and *Halticoptera* (Pteromalidae) are the most frequently mentioned. A series of experiments in California on chrysanthemums produced for cut flowers indicate good potential for biological control of leafminers using *Diglyphus intermedius* and *Chrysocharis parksi*. Other crops, such as gerbera and marigold, are likely candidates for inclusion in a leafminer biological control program. A computer program to assist in predicting parasite success against leafminers has been developed in California.

Whiteflies (Homoptera - Aleyrodidae)

Whiteflies are primarily tropical, but several of the approximately 1,100 species in the family have become pests in temperate areas as well as in greenhouses. The most important whitefly pest of protected floricultural crops is the greenhouse whitefly, *Trialeurodes vaporariorum*. This insect, first described in 1856, has caused problems on protected crops for more than 100 years and from all indications will continue on its present course. The greenhouse whitefly has been recorded on 249 plant species in 84 families. Another species, *Bemisia tabaci*, the sweet potato, tobacco or cassava whitefly, is also widespread and established as a pest of protected ornamental crops in North America. The sweet potato whitefly, first described in 1899, is generally considered to be a pest in more tropical areas, but may become more serious than the greenhouse whitefly simply because it feeds on more plant species. The sweetpotato whitefly attacks 506 plant species in 74 families. It is the most important virus vector among whiteflies. Another species, the bandedwinged whitefly, *Trialeurodes abutilonea*, has become more common in recent years. To date, this species hasn't reached outbreak levels on ornamental plants.

Biology

Adults and immature stages insert their piercing-sucking mouthparts into phloem tissue and remove plant fluids. Some females live up to two months and produce 30 to 500 eggs, but most probably live about one month and produce 60 to 80 eggs. Egg production in these insects sometimes increases with increasing adult whitefly density. Females usually begin egg production within several days of emergence. Although all the life cycle is very temperature-dependent, eggs generally hatch in seven to 10 days. The newly-hatched nymphs, or "crawlers," move for a short distance before settling down to feed. Following this, the immatures don't change location until adult emergence. After three molts a "pupal" stage is formed, from which adults emerge in about

*Adult greenhouse whitefly. White and dark eggs
are also visible.*

six days. Newly-emerged adults are pale green to yellow at first, but soon
become covered by a powdery, white wax. Adults have four wings. Whiteflies
complete their egg-to-adult cycle in 21 to 36 days, depending on temperatures.
Greenhouse whiteflies do best at moderate temperatures, while sweet potato
whiteflies (and bandedwinged whiteflies) do best at higher temperatures. All
growth stages are found on lower leaf surfaces. Adults usually prefer upper
leaves for feeding and egg-laying.

The whitefly species are usually easy to distinguish. Adult sweet potato
whiteflies are usually smaller, more yellow and more active than greenhouse
whiteflies. Sweet potato whiteflies normally hold their wings more vertically,
while greenhouse whiteflies hold wings nearly horizontally. Bandedwinged
whiteflies, as the name implies, have rather indistinct dark bands on the wings.
The immature stages of greenhouse and sweet potato whiteflies are also differ-

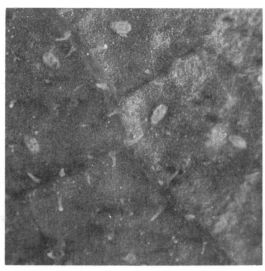

*First-instar greenhouse whitefly nymphs and
eggs on leaf underside.*

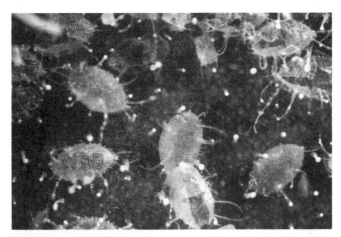

Greenhouse whitefly pupae.

ent, especially in the pupal stage. Sweet potato whitefly pupae are convex or flattened, without spines or fringes. Greenhouse whitefly pupae, on the other hand, have vertical sides with obvious spines and fringes of setae around the top perimeter. On some host plants (poinsettia) sweet potato whitefly immatures are yellow and have a "lemon drop" appearance.

Economic impact

Much of whiteflies' economic impact is due to their wide host plant range. On floricultural crops, a few whiteflies can be nuisance pests. The mere presence of the small, flying adults can result in consumer complaints. High populations weaken plants, and reduced plant vigor may eventually result in plant death. Whiteflies (and other piercing-sucking insects such as aphids, mealybugs and scales) also produce a sticky substance called honeydew, a digestive by-product rich in sugars. Often, enough honeydew is produced to act as a substrate for a black sooty fungus. If this fungus becomes well-established, leaves or even

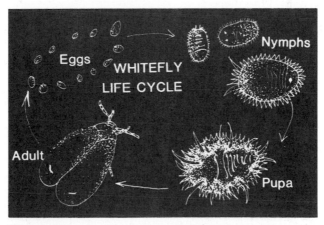

Whitefly life cycle. All stages occur on leaves.

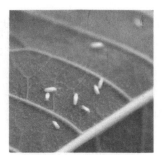

*Sweet potato whiteflies on
poinsettia leaf underside.*

entire plants can be killed because of reduced photosynthesis and respiration. Of course, the black fungus also makes plants unsightly. Other increasingly important impacts of whiteflies are as virus vectors. Most virus problems are associated with sweet potato whiteflies rather than greenhouse whiteflies. Although most virus problems resulting from whiteflies are on vegetable crops, cotton or soybeans, several whitefly-transmitted virus diseases occur on ornamental plants. These include diseases on pelargonium, hibiscus, rose, salvia, zinnia and anthurium. As sweet potato whiteflies continue to spread and cause problems, virus-caused plant diseases will also increase. See the discussion in Chapter 15.

Detection

Inspect whitefly-susceptible crops weekly. Be especially alert when receiving shipments of known, whitefly-host plants. Whiteflies don't occur uniformly throughout the crop, and there will be definite high number areas as well as areas without insects. Pay special attention to plants near open side ventilators and entry areas, but don't neglect plants in the center of the production area. Shake plants, turn leaves over or use a mirror to observe leaf undersides. A 10X to 15X magnification hand lens is needed to observe eggs and small nymph

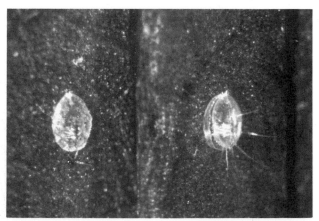

*Empty sweet potato whitefly (left) and greenhouse whitefly
(right) pupal skins.*
Photo by James R. Baker, North Carolina State University.

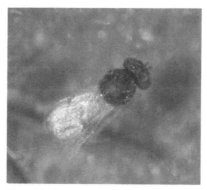

Encarsia formosa, *a parasite of greenhouse and sweet potato whitefly nymphs.*

stages. Yellow sticky traps placed throughout the crop will help detect adults, but traps won't replace plant inspection as a method of detecting low whitefly populations. Vertical traps, placed within the crop canopy next to the upper leaves, are satisfactory for detecting both sweet potato and greenhouse whiteflies. Sweet potato whiteflies are sometimes best detected by using yellow horizontal traps placed within the crop canopy or on the ground or greenhouse bench. Trap spacing can vary, but should be no more than 50 to 60 feet apart.

Cultural/physical control

Weed removal inside and outside the crop production area is very important. If possible, use screens and/or insect barriers to separate stock plant production areas from other production. On some outdoor crops, mulching with sawdust, straw or yellow polyethylene sheets reduces virus transmission by sweet potato whiteflies. The theory behind this is that the adults are attracted to the mulch by color and are killed after a short exposure to the mulch's high temperatures.

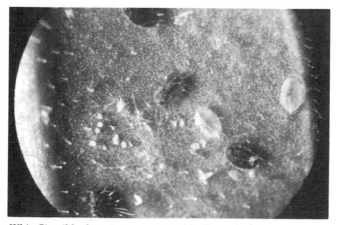

Whiteflies (black scales) parasitized by Encarsia formosa.

Apparently, this method works only for a short time when plants are small. In greenhouses, large numbers of yellow traps may be used to suppress populations, but only if the mass-trapping program is begun before a high whitefly population develops.

Physically removing lower leaves from plants containing large numbers of immature whiteflies helps reduce population increases. This process is labor intensive, but so are many pesticide applications. Obtaining adequate coverage on these leaves with pesticides is extremely difficult.

Pesticides

The rapid life cycle, high fecundity on favored host plants, location of the developmental stages on leaf undersides and pesticide resistance combine to make whitefly control difficult. Whitefly populations from different areas have been reported to be resistant to one or more organophosphate, carbamate and pyrethroid insecticides. The most effective control programs are based on preventing large population increases, so early detection is important.

The basic control procedures and pesticides are the same for both greenhouse and sweet potato whiteflies. After detecting a whitefly infestation, a standard recommendation is to make frequent applications (every five days) if using non-systemic materials. Maintain this application schedule for about 30 days before extending the treatment interval. In most cases, however, weekly applications for the same period usually provide adequate control.

Pesticide application technique is very important in whitefly control programs and must relate to the pesticide's mode of action. For contact insecticides, direct sprays toward leaf undersides and use the smallest spray drops possible. If necessary, adjust plant spacing for better spray coverage. Whenever possible, rotate spray applications with aerosol, fog or smoke generator applications.

Biological control

The primary natural enemy used against the greenhouse whitefly is the parasitic wasp *Encarsia formosa*. The adult parasite deposits an egg inside third-stage greenhouse whitefly nymphs, and the developing parasite kills the whitefly, turning it black in the process. Thus, it's easy to determine the biological control program progress by observing the percentage of black whitefly immatures on leaf undersides. The use of this parasite is widespread on protected tomato and cucumber crops in Europe and Canada but limited on floral crops. Poinsettia, gerbera and alstroemeria offer excellent possibilities for using this parasite. Once it becomes adapted to sweet potato whiteflies, *E. formosa* can be a very effective parasite against this species as well. Rather than turning black, parasitized sweet potato nymphs turn light brown. Experiments are underway in several locations to determine the best way to use *E. formosa* on ornamental plants. A predatory coccinelid beetle, *Delphastus pusillus*, has been very effective in reducing high sweet potato whitefly populations. This predator is available in limited quantities from commercial insectaries.

If parasites or predators are used, pesticide applications need to be markedly changed. Most biological controls, and especially Encarsia parasites, are extremely sensitive to insecticides. Some insecticides (methomyl, endosulfan, many pyrethroids) have effects that remain for more than one month and prevent Encarsia establishment.

Other biological controls for greenhouse whiteflies are fungi, including *Verticillium lecanii* and *Aschersonia aleyrodis*. Both fungi are applied against greenhouse whiteflies. Although *V. lecanii* is available in some European countries, no commercial formulations are available for greenhouse use in the United States. Results of commercial and experimental usage of *V. lecanii* have been mixed, and no firm guidelines for using this fungus are available. High humidities and/or water on leaves are necessary for fungi establishment. But once established, nearly complete control occurs in about two weeks. In addition to greenhouse whitefly control, strains of this fungus have been used to control aphids and thrips. No commercial formulation of *A. aleyrodis* is available at this time. This fungus, unlike *V. lecanii*, doesn't affect whitefly adults and doesn't produce an epizootic under greenhouse conditions. Thus, repeated applications with thorough coverage are necessary. The primary advantage of *A. aleyrodis* is that it may survive more easily in lower moisture conditions than *V. lecanii*.

Sweet potato whiteflies are also affected by similar fungi. A fungus, *Paecilomyces fumosoroseus*, has been very effective against sweet potato whiteflies (and other insects), and has been patented by the University of Florida. The proper environmental conditions for its use are being investigated.

Aphids (Homoptera - Aphididae)

There are more than 4,000 aphid species. Many aphid species are found on only one, or at most a few, host plants but some have very large host plant ranges. Several species can infest ornamental crops, but the most common ones are the green peach aphid (*Myzus persicae*), melon/cotton aphid (*Aphis gossypii*), chrysanthemum aphid (*Macrosiphoniella sanborni*), rose aphid (*Macrosiphum rosae*), potato aphid (*M. euphorbiae*), foxglove aphid (*Aulacorthum solani*) and leaf-curling plum aphid (*Brachycaudus helichrysi*). *M. persicae* is mentioned most often because of its worldwide distribution, very wide host plant range (more than 400 host plants), viral disease transmission (vectors more than 150 viruses or viral strains) and resistance to control. *A. gossypii* can also be very difficult to manage for many of the same reasons. For example, *A. gossypii* is known to transmit more than 50 viruses.

Biology

Aphids are soft-bodied, generally sluggish insects with piercing-sucking mouthparts, which are inserted into the phloem tissue of plants to remove fluids. Although often green, aphids can be many colors, including black, brown, pink, red or white. Aphids are the only insects that possess tubes, or siphunculi, on the abdomen. These tubes are often used by taxonomists to identify different species and sometimes make aphids appear as if they are jet-propelled. In greenhouses and tropical areas, all aphids are usually females that produce live young (nymphs). On outdoor plants in temperate zones, the life cycle includes the appearance of males and an egg stage that overwinters. Each female can produce about 50 to 250 nymphs during her life span, and

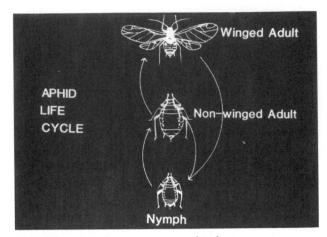

Aphid life cycle. All stages occur on the plant.

these nymphs can begin reproducing in seven to 10 days. Aphid reproduction depends, to a large extent, on host plant quality and nutrition. Of course, temperature is also very important. Adult aphids may be winged or wingless. Depending on population density and/or host plant condition, winged individuals sometimes form when aphids become overcrowded on their food plants. These winged individuals disperse within the crop and are often the first to arrive in a new location from outside the immediate area (from field crops or overwintering sites).

Economic impact
Usually, aphids are a nuisance because large numbers are required to actually affect growth of most plant species. The insects, or their white, cast-off exoskeletons, detract from a plant's value. Aphids produce honeydew, which can cover leaves and flowers with a sticky layer. If humidity conditions are right, a black, sooty fungus forms on the honeydew. Aphids are also important as vectors of plant viruses that affect many crops.

Adult green peach aphid and nymph.

Detection

Winged aphids are attracted to yellow sticky traps, so use the traps to monitor for aphids (as well as other pests). Many aphids don't have wings, however, so plant inspection is very important in detecting aphids. Pay special attention to terminal shoots and flower buds *before* flowers open. Another good way to detect aphids is to look for white skins left on plants during the molting process. Frequent plant inspection, especially of new shipments or after windy thunderstorms, also helps detect aphids. The presence of ants can be another clue to aphids being present. Some ant species "tend" aphids for their honeydew. In return, the ants protect aphids from predators and parasites.

Cultural/physical control

Cultural control includes removing weeds from around the plant production area. Weed host plants often serve as reservoirs for reinfestation. Fertilizer management is also very important since high nitrogen levels may promote higher aphid populations.

Physical control methods include screens or other barriers. Using screens is especially important in stock plant production areas to reduce the threat of viral transmission. Workers shouldn't wear yellow clothing because aphids attracted to the color may be carried into previously noninfested areas on this clothing.

Pesticides

Aphids can be control problems for several reasons. They have high reproductive capacities and often occur on lower leaf surfaces deep within plant canopies or in flowers. Pesticide resistance, especially among green peach aphid and cotton/melon aphid populations, is quite widespread. Individual aphid populations within each species vary widely in their susceptibility to certain pesticides. Many organophosphate and carbamate insecticides are no longer effective. The good news in control is that there generally are no dormant growth stages (eggs, pupae) to contend with, and all development stages occur on plants. Thus, if pesticide resistance isn't a problem, pesticide control should be straightforward and relate most to pesticide selection, application method and application interval. After discovering an aphid infestation, establish control by making two applications of an effective systemic insecticide about three to four

Adult winged green peach aphid.

Aphid skins on lily.

weeks apart. Apply nonsystemic insecticides three to four times at weekly intervals.

Biological control

Parasites, predators and fungi can potentially manage aphids on some ornamental crops. Most biological control programs are used on protected vegetable crops, but there is increasing interest in using biological controls on ornamental plants. Predators, such as larvae of the predatory midge *Aphidoletes aphidimyza* larvae and lacewing *Chrysoperla* spp. larvae have been successfully used. Both are available from commercial insectaries. Parasites in the genus *Aphidius* often occur naturally with minimal pesticide use, but usually too late to achieve control before high aphid populations develop. Parasitized aphids appear as light brown "mummies" on leaves. Both predators are available from commercial insectaries. The fungus *Verticillium lecanii* is effective in controlling some aphid species on chrysanthemums and other crops. A commercial product is available, but not in North America.

Aphid infected with Verticillium lecanii *fungus (left).*

Scale insects (Homoptera, families Diaspidae, Coccidae and Pseudococcidae)

Scale insects are a very large and diverse insect group. Worldwide there are about 6,000 species, with 1,000 species occurring in North America. All scale insects have piercing-sucking mouthparts and similar life histories. In all scale-insect families, females and males differ radically. Males of mealybugs usually resemble typical insects and have well-developed legs, eyes, antennae and, usually, wings. The three main families of scale insects show significant differences in biology and development. The three families are represented by the armored scales, soft scales and mealybugs.

Biology

The family Diaspidae are the so-called armored scales. This group gets its name from the adult's hard, waxy cover over its body. This covering is a mixture of wax and cast skins and comes in many shapes. It helps protect the insect from weather, parasites, some insecticides and protects the eggs produced by most species. Within the same species male and female coverings may vary in size and shape. The covering can be separated from the scale's body. Over time, extremely high populations can develop on plants.

As mentioned, most armored scale species females lay eggs that hatch beneath the covering into crawlers. The crawlers move to new plant growth where they settle and feed. At this time the females lose their legs and remain sessile for the rest of their lives. Females pass through three instars and the males, five. There may be as many as four generations per year. Armored scales don't produce honeydew.

The soft scales are in the family Coccidae. They are called soft scales because they don't form a scale similar to the armored scales until they mature. Unlike the armored scales, the soft scale covering can't be separated from the body.

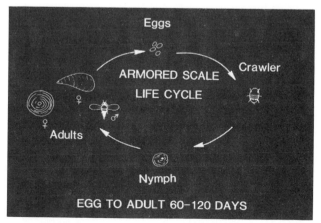

Armored scale life cycle. Males can fly, but all other stages occur on plants.

Heavy armored scale infestation on rose stem.

Soft scales are generally larger than armored scales, sometimes reaching a length of 3 to 4 millimeters. These insects' reproductive potential is tremendous. Soft scales may reproduce with or without mating, and females may lay eggs or produce live young. Females pass through three instars, and males, five. In warm climates and in greenhouses, there may be five or six generations per year. Soft scales produce large amounts of honeydew.

Mealybugs belong to the scale family Pseudococcidae and are the least scale-like of the group. They are soft-bodied insects that produce a waxy powder over their bodies. Most mealybugs also produce waxy projections around their bodies, and many have long filamentous "tails" consisting of waxy projections. A mealybug infestation sometimes appears as small pieces of cotton on the plant. This is especially true when females produce masses of waxy threads in which the eggs are laid. After eggs hatch, crawlers move about until they locate a suitable feeding site. Unlike other scale insects, mealybugs retain their legs throughout their development and move around and among plants. Once a suitable feeding site is located, however, movement is minimal. Preferred feeding sites include nodes and crotches of host plants. Females pass through four instars and males, five. Mealybugs often form large colonies around nodes and where stems branch. There may be as many as six generations per year in greenhouses and tropical areas. As with soft scales, mealybugs produce large amounts of honeydew.

Economic impact
This insect group has a very wide host plant range. Populations develop over relatively long time periods and high populations weaken and kill plants. Chlorosis and leaf drop can occur with lower populations. Honeydew deposition from soft scales and mealybugs covers leaves with a sticky coating, and if humidity is high, a black, sooty fungus grows on the honeydew. The cost of a control program can be high, and sometimes, replacing plants may be less expensive.

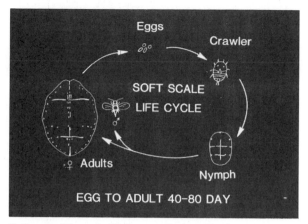

Soft scale life cycle. Males can fly, but all other stages occur on plants.

Detection

Early detection is the most important aspect of a management program for these insects. Scale insects can occur on nearly all plant parts. Pay particular attention to new plant shipments because infestations often begin on only a few plants and build up to large numbers over time. Separate any infested plants from other plants to reduce spread. Ants may help move scale insects throughout the production area. The presence of ants, wasps, bees or predatory insects should alert you that a scale insect infestation may be present. Because soft scales and mealybugs produce honeydew, look for sooty fungus on leaves.

Soft scales on stem and leaves.

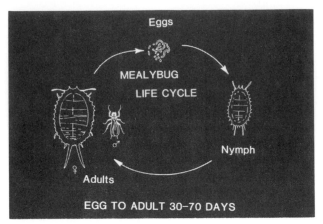

Mealybug life cycle. Males can fly, but all other stages occur on plants.

Pesticides

Many pesticides are effective only against crawlers. If the infestation is already well-established, overlapping generations (all stages) will be present. Apply nonsystemic pesticides at 14- to 21-day intervals. As many as six to eight applications may be required. Because these are piercing-sucking insects, systemic insecticides would seem the best kind to apply. Many infestations occur on older, woody plants, however, and systemic insecticides are usually most effective if plants aren't woody and are growing rapidly. Apply systemic materials every three to five weeks. Observe new growth to ensure that the infestation isn't spreading. Dried up and "flaky" insects are other indications of successful control.

Biological control

For soft and armored scale insects, both predators and parasites have been used. Two species of ladybird beetles, *Chilocorus nigritis* and *Lindorus lophanthae*

Adult female and immature mealybugs on leaf. The white, fuzzy material is an egg sac.

are commercially available for use in pest management programs. For mealy-bugs, the Australian ladybird beetle *Cryptolaemus montrouzieri* (the mealybug destroyer) has been used successfully on numerous occasions. This predator's larvae resembles mealybugs' larvae, and are a type of wolf in sheep's clothing. The common ladybird beetle, *Hippodamia convergens*, may provide some control if released in large numbers.

A parasite, *Metaphycus helvolus*, is being sold to manage several species of soft scales. Another parasite, *Aphytis melinus*, is used against some armored scales. For citrus mealybugs, a tiny wasp, *Leptomastix dactylopii*, has been used to supplement control obtained with predators.

Caterpillars (Lepidoptera)

Many Lepidoptera species attack floral crops. Most species are in the family Noctuidae, including cabbage loopers (*Trichoplusia ni*), beet armyworms (*Spodoptera exigua*), Egyptian cotton leaf worm (*S. littoralis*) and cutworms such as the variegated cutworm (*Peridroma saucia*). Other pests are in the families Pyralidae, including European corn borers (*Ostrinia nubilalis*) and Tortricidae, including the omnivorous leafroller (*Platynota stultana*). Most of these insects are rather general pests of varied agricultural and horticultural crops, inside and outside greenhouses. Their pest status is enhanced, as with other pests, by pesticide resistance and plant distribution. During the past few years, an Egyptian cotton leaf worm infestation occurred in Ohio and New York on ornamental plant cuttings from the Middle East. This is a quarantine pest in the United States, and caused significant financial losses to the greenhouses receiving those plant shipments.

Biology

The only common aspect of the above species is that the adults are rather dull-colored moths. The adults generally are most active at night or on dark days. In many cases moths may be attracted into greenhouses by lights used in

The mealybug destroyer, a ladybird beetle predator of citrus mealybugs.

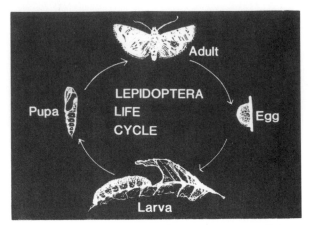

*Lepidoptera life cycle. Different stages may occur off or on
plants, depending on the species.*

plant production. Larvae all have chewing mouthparts and can cause conspicuous damage by consuming leaves and flowers, by tying or rolling leaves, boring
into stems and buds or cutting plants off at the base. The most common species
are described below.

Cabbage looper. These caterpillars are commonly found on chrysanthemum
foliage and flowers, particularly late in the summer in northern states. The
larvae are slightly longer than $1^{1}/_{4}$ inches when mature and are pale green with
white stripes on each side and along the back. Adult females lay eggs singly on
the foliage. After seven to 10 days, eggs hatch and the tiny larvae begin to feed.
Damage by young larvae causes a "window pane" appearance on leaves—the
leaves are not completely eaten through, and a thin, transparent cell layer
remains. Larvae develop for two to three weeks, consuming progressively more
plant tissue as they grow. They may eat entire leaves and flowers. Pupation
usually occurs on lower leaf surfaces, and the adult emerges in 10 to 14 days.

Cabbage looper.

Fully-developed beet armyworm larva.
Photo by James F. Price, University of Florida.

Beet armyworm. These are among major pests of chrysanthemums and other plants, particularly in southern states. Occasionally, moths may move north by flying or on infested plants, so that all areas are subject to this insect's attack. Eggs are laid in groups of about 100 on leaf undersides. The eggs hatch in two to nine days. Young larvae are often found feeding near growing points, often webbing young leaves together. Sometimes, feeding by small larvae will cause plants to become pinched, resulting in excessive branching, an infestation symptom. Older larvae will consume entire leaves and flowers and may bore into flower buds. The larvae are indistinctly striped, green-to-almost-black caterpillars that are about $1^1/4$ inches long when fully developed. A prominent dark spot is usually just behind the head area on older larvae. Larval development occurs in seven to 16 days, depending upon temperatures. Adults emerge from pupae in four to 11 days.

Leafroller. The larvae of this Lepidoptera family sometimes become problems on numerous ornamental plants. They are considered major rose pests, and have caused problems on poinsettia and geranium. Adults are small moths that

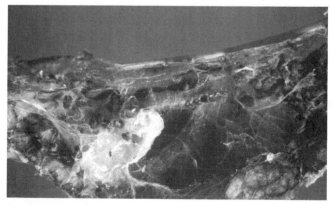

Leafroller larva.

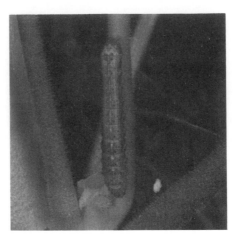

Variegated cutworm larva.

lay eggs in greenish clusters on host plant leaves. After eggs hatch in seven to nine days, the young larvae begin to feed on leaves. Initially, leaves are skeletonized or simply "gouged," but later the larvae web, tie or roll leaves together. This injury significantly affects plant appearance. The larval period lasts from 30 to 50 days, and pupation occurs on the plants. Adults emerge in about 10 days.

Cutworm. Cutworms are caterpillars that spend daylight hours beneath pots and/or growing media, feeding at night. Feeding damage may result in plants being cut off at the base, or leaves, buds and flowers may be damaged by so-called climbing cutworms. Therefore, damage may not be accompanied by a visible insect. One of the most common cutworms affecting ornamentals is the variegated cutworm, a climbing species. Adults generally lay several hundred eggs, which hatch in four to five days. Larval development takes 20 to 35 days. When fully developed, the larvae may be two inches long. Larval color varies, but generally is gray to black. There may be yellow or white stripes and spots. Cutworm infestations are frequently localized within a crop.

Economic impact

As described above, larvae can be very destructive by direct feeding damage. Early instar larvae often feed on terminals, causing stems to branch. Larger larvae will consume leaves and flowers and bore into stems. Of course, any infestations of quarantine significance can also be very serious. In the situation with the Egyptian cotton leaf worm described above, the greenhouses involved were shut down for many weeks until the infestation was certified as having been eliminated by inspectors from the U.S. Department of Agriculture.

Detection

Adults are attracted to blacklight traps, but placing traps inside a greenhouse may actually attract egg-laying adults. Light traps don't discriminate among species caught. This can make identifying economically important species very difficult. Pheromones (sex attractants) are very species-specific and are available for most economically important Lepidoptera. These traps aren't very successful inside a greenhouse, but they are effective in outdoor production areas. Knowing when adults are present in an area provides an early warning of a possible infestation and helps time insecticide applications.

Thorough plant scouting for plant injury and/or larvae is an important detection method. Young larvae often don't feed completely through leaves, resulting in "windows." Fecal pellets on the foliage are another infestation sign.

Pesticides

Several pests, notably the beet armyworm and cabbage looper, have developed resistance to one or more insecticides. Several pesticides, representing nearly all classes of compounds, are used in an effort to manage these pests. For example, pyrethroids were initially effective against beet armyworms, but reports of resistance soon followed. An important aspect in managing Lepidoptera is detecting the infestation before larvae become large. Smaller larvae are much easier to control. Systemic insecticides are generally not very effective against the larvae, so thorough coverage with contact/residual insecticides is required.

Biological control

The primary "biological" control agent in commercial use is the microbial insecticide *Bacillus thuringiensis* (B.t.). Actually, B.t. is not a true biological control, but often is placed in this category. This product is available under several brand names and formulations. Other controls, not yet used commercially on ornamentals, include different viruses that attack individual species.

Fungus gnats (Diptera - Sciaridae)

Adult fungus gnats are widespread insects found nearly everywhere plants are grown or maintained. One of the most common species is *Bradysia coprophila*, but other closely-related species also occur. The adults are often seen running across the growing medium surface. They can fly considerable distances,

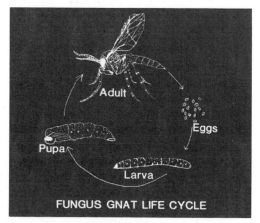

Fungus gnat life cycle. Egg, larva and pupa occur in the growing media.

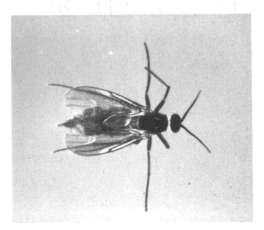

Fungus gnat adult.

though, and the adults are sometimes nuisances in homes, office buildings, hotels and retail flower shops.

Biology

The life cycle consists of egg, larva, pupa and adult. Adults are dusky, gray flies with long antennae, about ¼-inch long. Adults don't feed on plants. Females lay eggs in cracks and depressions in the growing medium surface. They are white and visible with a hand lens if lighting is adequate. Each female may lay more than 100 eggs. Two to four weeks are necessary to go from egg to adult. In five or six days the eggs hatch into white, translucent larvae with shiny black heads. The larvae pass through four instars and are about ¼-inch long when fully developed. They are easily visible on growing medium surfaces at this point. Most larvae occur in the top inch of the growing medium. Larvae feed on fungi, decaying organic matter and healthy plant tissue. After 10 to 14 days the larvae pupate within silky chambers in the growing medium. Several days later adults emerge. Two to four weeks are necessary to go from egg to adult.

Fungus gnat larvae.

Economic impact

As mentioned, the adults do no direct damage to plants, but are nuisances in many interior situations. Adults may carry plant pathogens from diseased to healthy plants. Direct injury to seedlings and transplants by larval feeding sometimes occurs. Plants with limited root systems may suffer stunted growth or death. Once the root system is well established, direct feeding injury is unlikely. Feeding injury may provide an entry point for root-infesting pathogens. Fungus gnats are usually most numerous during the first few weeks to two months after potting or planting.

Detection

Fungus gnat adults can be seen flying or walking near the growing medium surface. They are also attracted to yellow sticky traps. Only a few fungus gnat monitoring traps placed horizontally on the growing medium surface are required to detect these insects. Larvae are often visible on the growing medium surface, but are most often detected when examining roots of unhealthy plants and cuttings.

Cultural/physical control

Fungus gnats will produce higher populations in certain kinds of growing media (media containing hardwood bark or manure). Highest numbers are produced during the first generations, so controlling them early is important when producing crops in such media. Avoiding media with the above components won't eliminate fungus gnats, but will reduce numbers. Covering the medium surface with a layer of sand will discourage egg-laying by adults.

Pesticides

Most pesticide applications are directed at the larvae. Make applications as drenches or coarse sprays to the growing medium surface, and repeat them in three to four weeks. Aerosols, fogs or smoke generators (space treatments) are effective against adults. Repeat application every four to five days. Some growers also treat areas under benches with hydrated lime or copper sulfate.

Biological control

The microbial pesticide *Bacillus thuringiensis* H-14 (Gnatrol) has been very effective as a drench. Weekly applications give best results. The nematodes *Steinernema carpocapsae* and *S. feltiae* have been successfully applied, also as drenches, in some experiments. Predatory mites in the genus *Hypoaspis* have controlled fungus gnat larvae.

Shore flies (Diptera - Ephydridae)

Shore flies usually exist in the same conditions as fungus gnats and are often confused with them. The species found in greenhouses are in the genus *Scatella*. Adult shore flies are nearly all black, have reddish eyes, white spots on the wings and short antennae. They resemble fruit flies and leafminer adults. Adults sometimes gather in large numbers on surfaces of pots, flats and irrigation matting, wherever algae are found. The larvae are maggot-like and light tan in color. Both adults and brown larvae feed on algae. Direct plant injury is rare, although larvae have been reported to injure roots of plants grown in substrates

Shore fly adult.

such as rockwool. Large numbers of adults create a nuisance and also leave "fly specks" on leaves. Adults are attracted to yellow traps.

Control

Control shore flies by eliminating large areas of algae with either physical (irrigation mat covers) or chemical (bromine, quaternary ammonium salts) methods. Some pesticides may also be effective, but these insects don't seem as easy as fungus gnats to control in this way.

Spider mites (Acari - Tetranychidae)

The most important mite species affecting protected floral crops is *Tetranychus urticae*, the two-spotted spider mite. Another species, *T. cinnabarinus*, the carmine spider mite, also can cause problems on several crops. Spider mites are among the most serious ornamental crop pests wherever they are produced. Small size, rapid reproduction and pesticide resistance all contribute to their pest status.

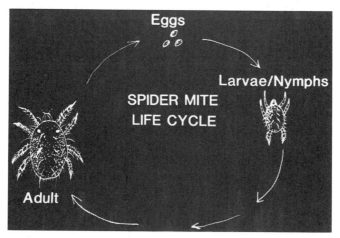

Spider mite life cycle. All stages occur on the plant.

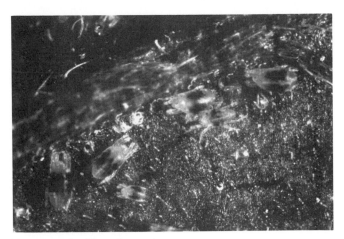

Two-spotted spider mites.

Biology

Both males and females exist in spider mite colonies, but females usually predominate. Adult females are about 0.5 millimeter long and range from light yellow or green to dark green, straw-brown and black. Normally, two dark spots are visible on either side of the abdomen. The carmine spider mite is naturally reddish and is commonly found on some plants such as carnation. Female *T. urticae* lay between 50 and 200 eggs, depending on the host plant. Development from egg to adult depends on temperature, host plant, plant tissue age and plant nutritional status, but the following will serve as a general guide. All development usually occurs on leaf undersides. Eggs hatch in four to seven days into six-legged larvae. The remaining stages are protonymph, deutonymph and adult. The total development time for the immature stages is

Spider mite feeding injury and webbing on palm.
Photo courtesy of National Foliage Foundation.

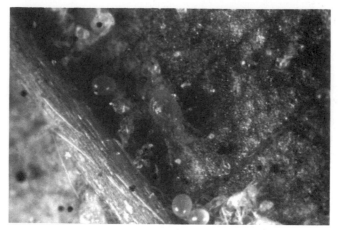

Spider mite predator, Phytoseiulus persimilis.

seven to 14 days. Spider mites develop extremely high populations during hot, dry conditions. Webbing is usually visible during moderate to heavy infestations.

In northern latitudes some spider mites will diapause (hibernate) in greenhouses during the winter months. In Ohio, for example, mites begin to diapause in November and emerge from diapause in mid-February. Diapausing spider mites are reddish and are sometimes confused with predators or the carmine spider mite. Diapausing mites do not feed and are less susceptible to pesticides. Also, not all spider mites in a greenhouse will diapause; some will remain active all year.

Economic impact
Heavy mite infestations can severely defoliate or kill plants. The mites penetrate leaf surfaces with stylet-like mouthparts and remove cell contents. Injured leaf tissue areas have no chloroplasts, giving the leaves a chlorotic, stippled look. Plant species vary widely in the amount of visible injury from a given number of spider mites. In addition to direct feeding injury, spider mites may inject toxins or growth-regulating chemicals, which can result in leaf dessication and defoliation. Spider mites produce large amounts of webbing. The webbing can cover foliage and flowers, markedly detracting from a plant's appearance.

Detection
Mites do not have wings, but can be blown on wind currents. Monitoring generally involves frequent inspections of plants, looking for feeding injury or webbing. Be especially alert during hot, dry periods and late in the production cycles on mite-susceptible crops. Carefully inspect new plant shipments before moving the plants into main production areas. Trap crops (plants more attractive to spider mites than the main crop) may also have some potential. A hand lens of 10X or 15X magnification will be very useful in detecting spider mites.

Cultural/physical control

Spider mites often develop higher populations on moisture-stressed plants. Thus, proper irrigation can help keep these high populations from developing. Using the old method of misting or syringing plants is also effective. Good air circulation helps reduce plant diseases. If predators are used for control (see below), the slower dispersal of spider mite populations in wet conditions will be an advantage. High nitrogen is also associated with severe mite outbreaks. Ensuring that no weeds are growing within the production area or immediately around it is another pest management tool. Keep workers from moving through known spider mite-infested areas because mites often disperse by clinging to clothing. Visit known, mite-infested areas last.

Pesticides

Frequent pesticide applications often help maintain low spider mite numbers throughout a crop production. On continuous crops, for example, rose, make applications two to three weeks apart--up to 25 each year. Despite this schedule, high mite populations often develop. With most pesticides, make at least two applications, about five days apart, to significantly reduce spider mite population. Use proper application techniques, because most mites will be found on leaf undersides. Be careful with pesticide selection. Certain pyrethroid insecticides can stimulate spider mite reproduction and/or cause the mites to become more active and form colonies in more crop areas as they disperse.

Biological control

Although predatory mites such as *Phytoseiulus persimilis* and *Metaseiulus occidentalis* are used on many protected food crops, the heavy pesticide application regimen normally followed on floral crops usually prevents the use of predators. On some crops, such as rose, chrysanthemum, gerbera and many foliage crops, there are possibilities for using predatory mites. Current studies are underway to develop practical, ornamental crop management programs. Predators with some pesticide resistance are available, but this resistance doesn't extend to pyrethroid insecticides. Many insecticides remain harmful to predators more than one month after application.

Other insect, mite and associated pests

Plant bugs (Hemiptera - Miridae)

Plant bugs represent the true bugs, and several species can cause problems. The tarnished plant bug (*Lygus lineolaris*) and four-lined plant bug (*Poecilocapsus lineatus*) are among the most common species. The tarnished plant bug is a light brown insect with lighter markings, and is about 1/4-inch long. Four-lined plant bugs are yellow with four black stripes on the wing covers and are slightly larger than tarnished plant bugs. Several other related species can also occur.

Plant bugs have piercing-sucking mouthparts. Females insert eggs into plant tissue singly or in small clusters. The eggs hatch into nymphs, which are very active and do considerable feeding. After four or five nymphal instars, the adults are formed. Adult plant bugs are winged. A generation can be completed in about 30 days or less, depending upon temperatures.

Tarnished plant bug.

This insect group feeds on a wide range of host plants. In the South they are often serious pests of field- or saran-grown ornamentals. In northern areas they may move into greenhouses late in the summer, when outdoor crops deteriorate. Plant bugs inject toxic saliva during feeding, resulting in dark spots on host plant leaves. The tarnished plant bug causes plants to branch excessively by feeding on growing points. Feeding on buds causes flowers to abort. Often, the damage caused by these insects appears before the pests are detected. In areas prone to plant bug injury, thorough crop scouting, particularly when close to flowering, is very important.

Beetles (Coleoptera)

Beetles are abundant insects, ranging in size from almost microscopic to very large. Adults have the front pair of wings hardened, forming a "shell" (elytra) over the rear pair. This insect group has chewing mouthparts. Many adults, including the Japanese beetle, cause severe foliage and flower injury through feeding. Beetle larvae are called grubs. Many species, for example the black

Beetle larva (white grub).

vine weevil and the group called "white grubs," including larvae of Japanese beetles, feed on plant roots. Various plant quarantines affect this group.

Detection usually occurs when the adult visibly feeds on plants or when symptoms associated with root injury appear. Traps that chemically attract adult Japanese beetles could be useful in determining when adult activity is occurring in any particular area. Traps aren't used for control. Adults can be killed with foliar sprays of several insecticides, but most effective control is usually directed at the larvae. Soil drenches or granular insecticide applications are necessary. Some uses of the microbial pesticide *Bacillus popillae* have been successful, as have nematodes applied to the soil.

Grasshoppers (Orthoptera - Acrididae)
Grasshoppers are rather general feeders and can cause locally severe injury to ornamental plants in many North American areas. They usually are most numerous and damaging in areas with average yearly precipitation between 10 and 30 inches. High populations generally develop in an area over a period of three or four favorable years, and individuals move into greenhouses from adjacent field crops.

Grasshoppers have chewing mouthparts and can consume entire plants or specific plant parts. Several insecticides will control these insects, but often only after considerable feeding injury has occurred. Place screens on vents and fan intakes to keep these large insects out of a greenhouse.

Rose midge (Diptera - Cecidomyiidae)
This insect pest *Dasineura rhodophaga* has caused problems on roses for many years. It was a serious pest of greenhouse roses from the late 1880s until the widespread use of DDT relegated it to nonpest status; however, severe injury has been reported during recent years. When present, it's one of the worst rose pests. Severe injury can occur with amazing speed. The adult is a small, dark, fragile fly, about $1/20$-inch long. It deposits eggs in groups of 20 to 30 in bud folds or on new terminal growth. The larvae are whitish maggots that feed on flower buds and terminals, preventing flowering. Infested shoots often are crooked and rose production stops completely. Pupation takes place on the growing medium. Rose midges complete a generation in about two weeks.

Control is difficult, often because when damage is obvious, the pests are no longer present. Most success has been obtained by directing pesticide applications at the growing medium to help kill larvae that drop off plants, or newly-emerged adults. Cultural control by pruning infested buds and tips and destroying them will also help.

Springtails (Collembola)
Many authorities don't consider springtails insects. Springtails are closely-related, however, and very common creatures not often seen because of their small size (2 to 3 millimeters) and their location in concealed situations. On potted plants they are often noticed when large numbers emerge from the growing medium after irrigation. Most species have a forked structure (furcula) folded forward under the abdomen and held in place by a clasping structure (tenaculum). These insects jump when the furcula is released from the tenacu-

Springtail.

lum against the substrate. Some species can jump up to 3 inches. This jumping ability can be used to distinguish between springtails and symphylans. Springtails have stylet-like mouthparts and feed on decaying plant material, algae, pollen, fungi, bacteria and young roots. Most species don't cause economic injury. Serious root damage, however, can occur on cut flower crops produced on ground beds. Although only necessary in a few instances, control is accomplished by applying insecticides to the soil.

Tarsonemid mites (Acari - Tarsonemidae)

The two best-known members of this group are the cyclamen mite *Stenotarsonemus pallidus* and broad mite *Polyphagotarsonemus latus*. The general life cycles of both species are similar. In contrast to spider mites, these tarsonemid mites do best in cooler, moister conditions. The tiny adults are less than 0.3 millimeters long and are colorless or tinted brown. Female cyclamen mites lay eggs on upper leaf surfaces, whereas broad mites lay eggs on leaf undersides and/or dark, moist places on plants. After two to 11 days, eggs hatch into

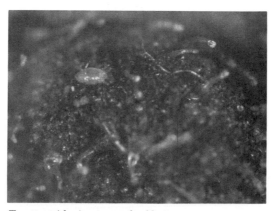

Tarsonemid mite among leaf hairs.

Lily bulb mite.
Photo by Mark E. Ascerno,
University of Minnesota.

whitish larvae that develop for three to seven days and then pass through a quiescent stage before molting into adults. The general time from egg to adult for cyclamen mites is 18 days and for broad mites, 10 days. Favorite host plants for cyclamen mites include African violets, ivy, snapdragon, geranium, cyclamen, azalea and fuchsia. Broad mite hosts include ivy, impatiens and peperomia. Injury symptoms may include leaf distortion, stunting, bronzing and plant death.

Injury caused by these tiny mites is the best way to recognize an infestation. The symptoms resemble those caused by pesticide injury or nutritional problems, but tend to be localized, rather than occurring everywhere in the crop. If only a few plants are affected, the best way to deal with the problem is to remove the plants. Spray larger affected areas with an effective pesticide. Several applications at weekly intervals will be required.

Bulb mites (Acaridae)

These mites, *Rhizoglyphus* spp., infest bulbs of many plant species (onion, narcissus, hyacinth, tulip and lily). The lily bulb mite *R. robini* is one of the most frequently encountered bulb mite species, simply because of Easter lilies' popularity. Lily bulb mites infest field-grown lily bulbs in the western United States and are often shipped on harvested bulbs to all areas of the country. This species is a secondary pest and prefers to feed on previously weakened plant tissue. They can invade healthy tissue, however, and sometimes reach high numbers on bulbs during the greenhouse production season.

Bulb mites' life cycle is as follows: Eggs are laid singly or in groups on bulb surfaces, near injured and decaying tissue or between bulb scales. Each female can produce about 100 eggs. The egg-to-adult cycle is completed in about 10 days at 80 F. Adult bulb mites are about 1 millimeter long, quite large for mites. They have a pearly white body and short, reddish legs. Bulb mites move slowly.

The economic injury caused by bulb mites varies. The exact relationship between bulb mite populations and plant injury through feeding damage or secondary invasion of plant pathogens (Pythium, Rhizoctonia, Fusarium) at the

*Lily showing injury associated with bulb
mite infestations.*
Photo by Mark E. Ascerno, University of Minnesota.

feeding sites isn't known. It is certain, however, there is some kind of relationship and treatment for bulb mites on a preventive or "insurance" basis is
usually suggested. According to information from the University of Minnesota,
keeping bulb mite numbers low will improve root rot control with fungicide
drenches. The best control for lily bulb mites is to soak bulbs in dicofol
(Kelthane 35WP, 1^1/$_3$ pounds per 100 gallons) for 30 minutes before potting.
Monthly fungicide drenches should also be applied.

Symphylans (Symphyla)
Symphylans are not insects but taxonomically are placed somewhere between
centipedes and insects. Adults are about 8-millimeters long, with 12 legs and
14 segments and less than 1 millimeter wide. Symphylans have long antennae
with about 60 segments. There are no eyes, and the antennae serve as sensory
organs. The adults move rapidly and attempt to escape from light.

Eggs, 0.5 millimeter diameter, are laid in the soil, about 30 centimeters deep.
They hatch in one to three weeks into tiny (1.5 millimeters) larvae. The larvae
are very sluggish and have six antennal segments. They are sometimes confused with Collembola (springtails), but symphylans cannot jump. The larvae
become very active after the second molt. Symphylans reach sexual maturity
in 40 to 60 days and may live more than two years.

Symphylans do best in cool, moist conditions, with the optimal temperature
about 62 F. As temperatures increase, they move to the subsoil, sometimes

Symphylan.
Photo courtesy of Oregon State University, Department of Entomology
Extension Service.

going down 2¹/₂ to 3¹/₂ feet. Because they aren't able to create their own burrows in the soil, symphylans use trails made by other organisms.

Symphylans injure plants by eating root hairs, chewing cavities in larger roots and stems and hollowing out seeds. In addition to cutting off the plant's food supply, the feeding injury provides a place for pathogens to enter the plant. Most symphylan injury occurs along paths and walkways not disturbed by cultivation or not pasteurized.

Detection usually occurs by eliminating other possibilities for poor plant growth. Wilting plants during sunny afternoons indicate possible root problems. One of these problems, with plants produced in ground beds in soil, could be symphylans. Place soil samples in a container, and add water slowly to force symphylans to the soil surface.

Control is based on prevention (avoid contaminated soil, use soilless media or produce plants on raised benches), cultural practices (deep cultivation) or pre-plant soil treatments (fumigation or insecticides). Many insecticides most effective for symphylan control have been, or soon will be, severely restricted because of groundwater concerns.

Slugs (Mollusca)

Slugs feed on a wide range of plants inside and outside greenhouses. Several species can occur. Slugs have caused problems on ornamental plants for many years. They are classified as Molluscs, which includes snails, clams and squids. Slugs range in length from less than ¹/₂ inch to more than 10 inches. Most species that affect ornamentals are about 3/4- to 1-inch long. Slugs are generally active at night and are found beneath flats, pots and dense foliage.

Slugs are essentially snails without shells. Eggs are laid in groups of 15 to 50, usually in cracks in soil or in loose soil. If soil moisture and temperature conditions are favorable, the eggs hatch. If not, eggs await those conditions. Slugs mature and begin to reproduce in three to five months. There may be three generations a year. Cool, wet conditions are best for slug survival and development. They are most active at night or on cool, cloudy days.

Slugs can feed and survive on nearly any kind of plant material, but have definite preferences if presented with a choice. Damage occurs through direct

Slugs.

feeding injury. Characteristic injury consists of irregular holes eaten in foliage. Sometimes, entire plants can be stripped of leaves. Usually, a shiny, slimy trail of mucous-like substance is present.

A good detection method is to place boards or other flat objects on the soil surface in the evening and look under them the following day. This may reveal slugs seeking moist, dark, daytime hiding sites. Shallow dishes of beer placed in the soil so that the top of the dish is flush with the soil surface will trap slugs. Achieve control by applying specialized slug baits containing a pesticide. Eliminate slug habitats through sanitation, raising flats or pots off the floor and eliminating excessive moisture.

Chapter 17

Pesticides for Insect and Mite Control

*I*ncluded in this chapter are the latest pesticide registrations for insect and mite control on protected ornamental plants. Because individual states or provinces may have additional (or fewer) registered materials, it's always a good idea to consult pesticide labels applicable to your area. Relatively few new insecticides and acaricides have been registered in the past few years, and prospects are uncertain. As described in Chapter 18, many materials that have been registered are in the "soft" pesticide category. This doesn't decrease their effectiveness, but does require the applicator to become educated concerning their characteristics and how best to use them. The slowdown in new registrations, the disappearance of older products from the market, plus pesticide resistance will create problems for ornamental plant producers and consumers.

Pesticide resistance is, and will continue to be, a major factor influencing pest management programs. Studies are underway to determine how best to deal with resistance, but it appears safe to say that the problem will get worse before it gets better. Growers simply have to look at the severe damage and associated costs due to leafminer (*Liriomyza trifolii*), western flower thrips (*Frankliniella occidentalis*), aphid (*Myzus persicae)* and whitefly (both *Trialeurodes vaporariorum* and *Bemisia tabaci*) infestations. Many problems with these pests are due to pesticide resistance. Growers should periodically rotate chemical classes to try to delay pesticide resistance. Change classes after each three or four applications (minimum of two applications). Avoid combinations or tank mixes of insecticides if possible, but sometimes this is the only way to control a particular pest.

The following table summarizes the registered pesticides for ornamental crops, pesticide formulations and application methods. Nearly all pesticides listed are registered for both greenhouse and outdoor use. Those pesticides registered only for outdoor use will be noted in the "application method" column. Including or omitting a product or formulation here doesn't constitute an endorsement or criticism. The chart is informational only. Individual pesticides and their chemical classifications are discussed following the chart.

General summary of insecticides and miticides used on ornamental crops in the United States to control major pest groups[1]

Insect or mite	Common name	Brand names	Application method[2]
Aphids	Acephate	Orthene	HV
		PT 1300	A
	Bifenthrin	Talstar	HV
	Chlorpyrifos	Dursban	HV
	Cyfluthrin	Decathlon, Tempo	HV, LV
	Diazinon	Diazinon	HV (outdoor use)
		Knox-Out	HV
	Dichlorvos	DDVP, Vapona	F, S
	Dimethoate	Cygon	HV (outdoor use)
	Disulfoton	Di-Syston	G (outdoor use)
	Endosulfan	Thiodan	HV, S
	Fenpropathrin	Tame	HV
	Fluvalinate	Mavrik	HV, LV
	Insecticidal soap	M-Pede, Safer's	HV
	Horticultural oil	Sunspray	HV
	Kinoprene	Enstar	HV
	Malathion	Cython, Malathion	HV (outdoor use)
	Methomyl	Lannate	HV
	Naled	Dibrom	S, V
	Nicotine sulfate	Black Leaf 40	HV
		Nicotine	S
	Oxamyl	Oxamyl	G
		Vydate L	HV
	Sulfotepp	Dithio	S
Caterpillars (Lepidoptera)	Acephate	Orthene	HV
		PT 1300	A
	Bacillus thuringiensis	Dipel, Victory	HV, LV
	Bifenthrin	Talstar	HV
	Carbaryl	Sevin	HV (outdoor use)
	Chlorpyrifos	Dursban	HV
	Cyfluthrin	Decathlon, Tempo	HV, LV
	Diflubenzuron	Dimilin	HV
	Fluvalinate	Mavrik	HV, LV
	Methomyl	Lannate	HV
	Permethrin	Pounce, Pramex	HV

Fungus gnats			
(adults)	Resmethrin	PT 1200	A
(larvae)	*Bacillus thuringiensis* H-14	Gnatrol	D
	Oxamyl	Oxamyl	G
	Diazinon	Knox out	D
	Kinoprene	Enstar	D
Leafminers	Abamectin	Avid	HV, LV
	Chlorpyrifos	Dursban	HV
	Cyromazine	Citation	HV
	Dichlorvos	DDVP, Vapona	F
	Dimethoate	Cygon	HV (outdoor use)
	Neem	Margosan-O	D, HV
	Nicotine sulfate	Black Leaf 40	HV
		Nicotine	S
	Oxamyl	Oxamyl	G
	Permethrin	Pounce	HV
		Pounce, Pramex	HV
Mealybugs, scales	Acephate	Orthene	HV
		PT 1300	A
	Bendiocarb	Dycarb	HV
		Ficam	HV
		Turcam	HV
	Bifenthrin	Talstar	HV
	Chlorpyrifos	Dursban	HV
	Cyfluthrin	Decathlon, Tempo	HV, LV
	Diazinon	Knox-Out	HV
	Dichlorvos	DDVP, Vapona	F, S
	Dimethoate	Cygon	HV (outdoor use)
	d-Phenothrin	Sumithrin	HV
	Insecticidal soap	M-Pede, Safer's	HV
	Naled	Dibrom	V
	Kinoprene	Enstar	HV
	Oxamyl	Oxamyl	G
		Vydate L	HV
Plant bugs	Acephate	Orthene	HV
(e.g. tarnished	Bifenthrin	Talstar	HV
plant bug)	Carbaryl	Sevin	HV (outdoor use)
	Chlorpyrifos	Dursban	HV
	Cyfluthrin	Decathlon, Tempo	HV, LV
	Fluvalinate	Mavrik	HV, LV
	Insecticidal soap	M-Pede, Safer's	HV

Mites			
(cyclamen mite)	Endosulfan	Thiodan	HV
	Dicofol	Kelthane	HV
(broad mite)	Dienochlor	Pentac	HV
(two-spotted spider mite)	Abamectin	Avid	HV, LV
	Bifenthrin	Talstar	HV
	Dichlorvos	DDVP, Vapona	F, S
	Dienochlor	Pentac	HV
	Fenbutatin-oxide	Vendex	HV
	Fenpropathrin	Tame	HV
	Fluvalinate	Mavrik	HV, LV
	Insecticidal soap	M-Pede	HV
		Safer's	HV
	Horticultural oil	Sunspray	HV
	Naled	Dibrom	V
	Oxamyl	Oxamyl	G
	Oxythioquinox	Joust, Morestan	HV, LV (outdoor use)
Slugs	Metaldehyde	Bug-Geta	B
		Deadline	B
		Slugit	B
		Snarol	B
	Methiocarb	Grandslam,	HV, B
		Mesurol, PT1700	A
Springtails	Malathion	Malathion	SSP
Thrips	Acephate	Orthene	HV
		PT 1300	A
	Bendiocarb	Dycarb	HV
		Ficam	HV
		Turcam	HV
	Chlorpyrifos	Dursban	HV
	Cyfluthrin	Decathlon, Tempo	HV, LV
	Diazinon	Diazinon	HV (outdoor use)
		Knox-Out	HV
	Dichlorvos	DDVP, Vapona	F, S
	Disulfoton	Di-Syston	G (outdoor use)
	Endosulfan	Thiodan	HV, S
	Fluvalinate	Mavrik	HV, LV
	Methomyl	Lannate	HV
	Malathion	Malathion	HV (outdoor use)
	Naled	Dibrom	V
	Nicotine sulfate	Black Leaf 40	HV
		Nicotine	S
	Resmethrin	PT 1200	A
		SBP-1382	HV
	Sulfotepp	Dithio	S

Whiteflies	Acephate	Orthene	HV
		PT 1300	A
	Bifenthrin	Talstar	HV
	Cyfluthrin	Decathlon, Tempo	HV, LV
	Dichlorvos	DDVP, Vapona	F, S
	Dimethoate	Cygon	HV (outdoor use)
	Disulfoton	Di-Syston	G (outdoor use)
	d-Phenothrin	Sumithrin	HV
	Endosulfan	Thiodan	HV, S
	Fenpropathrin	Tame	HV
	Fluvalinate	Mavrik	HV, LV
	Insecticidal soap	M-Pede	HV
		Safer's	HV
	Horticultural oil	Sunspray	HV
	Kinoprene	Enstar	HV
	Methiocarb	PT 1700	A
	Methomyl	Lannate	HV
	Naled	Dibrom	V, S
	Neem	Margosan-O	HV
	Oxamyl	Oxamyl	G
		Vydate L	HV
	Oxythioquinox	Joust, Morestan	HV, LV (outdoor use)
	Permethrin	Pramex	HV
	Pyrethrins	Pyrenone	HV, LV
		Pyreth-it	HV, LV
	Resmethrin	PT 1200	A
		SBP-1382	HV, LV
	Sulfotepp	Dithio	S

[1]Before purchasing and using any pesticide, check all labels for registered use, rates and application frequency. Common and brand names used in the United States are given for all pesticides.

[2]Application methods: A = aerosol; B = bait; D = drench; F = fog; G = granules; HV = high volume spray; LV = low volume (concentrate) spray; S = smoke generator; SSP = soil surface spray; V = vaporize off hot pipes or hot plates.

Insecticides and miticides registered on ornamental plants

Following is a brief discussion of the pesticides mentioned in the preceding chart. For detailed information on these materials, contact your local or regional extension specialist.

Abamectin (Avid), class: natural product. Abamectin, registered for control of leafminers and spider mites, is an insecticide/miticide derived from soil microorganisms. It has also been effective against whiteflies, thrips and some aphid species. Abamectin is not a "true" systemic pesticide, but has translaminar activity in some plants, meaning it moves from upper to lower leaf surface.

Acephate (Orthene, PT 1300), class: organophosphate. Acephate is a systemic insecticide effective against a wide range of chewing and sucking pests. It is an excellent insecticide, but has caused phytotoxicity on several chrysanthemum cultivars and foliage plants. Be careful when applying this material to any new cultivar. Wait two weeks for symptoms to appear.

Bacillus thuringiensis var. Kurstaki (Dipel), class: microbial. A microbial insecticide, *Bacillus thuringiensis* or simply B.t. is effective against several species of chewing caterpillars. The insect must eat this material for it to be effective, so thorough coverage is necessary for best results. Caterpillars stop eating soon after consuming foliage containing residues of B.t. but may not die for two to three days. When B.t. is used as directed, no plant injury has been reported. This material is compatible with natural enemies for biological control of other pests.

Bacillus thuringiensis H-14 (Gnatrol), class: microbial. This microbial insecticide is a different strain of the B.t. described above. It has particularly good activity against certain fly larvae and has given good control of fungus gnat larvae. This formulation has the same low mammalian toxicity and plant safety as the B.t. formulation listed above and is compatible with natural enemies.

Bendiocarb (Dycarb, Ficam, Turcam), class: carbamate. Bendiocarb is an insecticide registered for control of several common insect and mite pests on a wide range of ornamentals. It has been especially effective in controlling mealybug and scale insects when applied as a foliar spray. As a drench it's registered for control of black vine weevil larvae but also kills fungus gnats. Avoid using alkaline water for maximum effectiveness. Apply sprays carefully to limit runoff, as root injury has occurred on several foliage plant species.

Bifenthrin (Talstar), class: pyrethroid. Talstar is an insecticide/miticide effective in controlling some species of aphids, spider mites, whiteflies, mealybugs, scales and caterpillars found on floral and foliage crops. Talstar acts through contacting pests and has no known systemic or vapor activity, so thorough coverage is necessary to obtain good control.

Carbaryl (Sevin), class: carbamate. Carbaryl is used against a wide range of insects on outdoor ornamentals. Available in several formulations, it suits various needs. Residual activity is short, so repeat applications are necessary.

Chlorpyrifos (Dursban), class: organophosphate. Dursban is a broad spectrum insecticide registered on most ornamentals. It's effective against aphids, mealybugs, scales, caterpillars and leafminers. Only the 50WP formulation is registered on greenhouse crops, but an EC formulation can be used on outdoor ornamentals. A microencapsulated formulation may also be available soon and should reduce this material's plant damage potential.

Cyfluthrin (Decathlon, Tempo), class: pyrethroid. Cyfluthrin is an advanced pyrethroid with broad-spectrum activity against insects and mites affecting ornamental plants. It controls whiteflies, scales, thrips, some aphid species and caterpillars.

Cyromazine (Citation), class: insect growth regulator. Citation is active primarily against leafminer larvae. This material has systemic activity and has been successfully applied as a drench, although current recommendations are to make foliar applications. When used as a drench, it has provided some fungus gnat control.

Diazinon (Knox-Out), class: organophosphate. Diazinon is a broad-spectrum insecticide with uses against many chewing and sucking pests. A microencapsulated formulation is used in greenhouses, while wettable powder and liquid formulations are available for outdoor use.

Dichlorvos (DDVP, Vapona), class: organophosphate. Dichlorvos is a vapor-active insecticide/miticide. It can be applied by vaporizing off hot heat pipes or hot plates or by fog, smoke generator or mechanical mist. It may injure some ornamentals, particularly if greenhouse temperatures are too high. Dichlorvos is effective at 60 to 65 F.

Dicofol (Kelthane), class: chlorinated hydrocarbon. Kelthane is registered as a foliar spray for two-spotted spider mite control. It's available in wettable powder and flowable formulations. Repeat applications often during warm weather.

Dienochlor (Pentac), class: chlorinated hydrocarbon. Pentac is registered for two-spotted spider mite control. Because Pentac is very slow acting, more than five days may be required for any noticeable activity. Repeat applications often during warm weather.

Diflubenzuron (Dimilin), class: insect growth regulator. Dimilin has only a limited registration on field- and greenhouse-grown chrysanthemums for beet armyworm control. This material actively interferes with larval growth and molting process.

Dimethoate (Cygon), class: organophosphate. Dimethoate is a broad-spectrum, systemic insecticide used on many outdoor ornamentals. Phytotoxicity can be a problem, so check the label for precautions.

Disulfoton (Di-Syston), class: organophosphate. Disulfoton is a granular, systemic insecticide/miticide with activity against several sucking pests. It's usually applied at planting, or shortly thereafter, for best results. Di-Syston is a toxic material registered only on outdoor ornamentals.

d-Phenothrin (Sumithrin, PT 1400), class: pyrethroid. This is an early-generation pyrethoid registered for whitefly, aphid, spider mite and mealybug control on a wide range of plants. Repeat applications are necessary to maintain control.

Endosulfan (Thiodan), class: chlorinated hydrocarbon. Thiodan WP, EC and smoke generator formulations are available. This material has been used for many years on a wide range of ornamental plants to control whiteflies, aphids, caterpillars and certain tarsonemid mites (cyclamen mite).

Fenbutatin-oxide (Vendex), class: organotin. Vendex is a miticide registered for use on a wide range of ornamentals. It's best used in a preventive control program rather than in an attempt to reduce an already heavy mite population. Available in wettable powder and flowable formulations, Vendex can be useful in an integrated pest management program that uses predatory mites for spider mite control, because the material is less harmful to predators than to spider mites. Thus, it can be used to regulate spider mite numbers.

Fenpropathrin (Tame), class: pyrethroid. Tame is an insecticide/acaricide active against whiteflies, aphids, mealybugs, scales, caterpillars and spider mites. It's one of the so-called advanced generation pyrethroid insecticides.

Fluvalinate (Mavrik), class: pyrethroid. Mavrik is an insecticide/miticide that has given fair to good control of many aphid, thrips and spider mite populations. Mavrik may generally be used safely on open flowers.

Horticultural oil (Safer's Sunspray), class: spray oil. Dormant oils aren't new in horticulture, but oils such as Sunspray can be applied to growing crops with little phytotoxicity. Oils have given good control of whiteflies, aphids and spider mites.

Insecticidal soap (M-Pede, Safer's), class: soap. Safer's Insecticidal Soap contains potassium salts of fatty acids. When properly applied, it's effective for control of aphids, mealybugs, whiteflies and spider mites.

Kinoprene (Enstar), class: insect growth regulator. Enstar has recently reappeared on the market in the United States after several years' absence. A growth regulator of low mammalian toxicity, it's effective against whiteflies, mealybugs, scales and aphids. Enstar is also quite "soft" on beneficial insects and mites.

Malathion (Cythion, Malathion), class: organophosphate. In greenhouses, Malathion is registered for use as a soil surface spray for millipede, springtail and sowbug control. Outdoors it has many uses as a foliar spray.

Metaldehyde (Bug Geta, Deadline, Slugit, Snarol). Metaldehyde is an old ingredient used for snail and slug control. It's available in several formulations, including pellets and liquid. Apply to the growing medium surface around the plants to be protected.

Methiocarb (Grandslam, Mesurol), class: carbamate. Grandslam is applied as a foliar spray for aphid, mite, slug and snail control. Mesurol is a pelletized bait formulation for slug control.

Methomyl (Lannate), class: carbamate. Methomyl is a broad-spectrum insecticide with activity against such diverse pest groups as caterpillars and whiteflies. It is registered on many outdoor ornamentals, but only a few states have registrations for greenhouse use.

Naled (Dibrom), class: organophosphate. Dibrom is registered for vapor or fog treatments. Pests controlled include spider mites, whitefly adults, aphids, leafrollers and mealybugs. Excellent control of the western flower thrips has also been reported. The concentrate is highly corrosive to metals.

Neem extract (Margosan-O), class: natural product. Margosan-O is a formulation containing three grams per liter of azadirachtin, a primary insecticidal compound in neem seed extract. This material is effective against whiteflies (immatures), leafminer larvae, as well as several caterpillar species. As with other insect growth regulators, activity is slow and results may not be apparent for several weeks.

Nicotine (Black Leaf 40, Nicotine Smoke), class: botanical. The smoke generator formulation of this vapor and contact-active insecticide is quite effective against aphids on many ornamental plants.

Oxamyl (Oxamyl, Vydate), class: carbamate. Oxamyl is a systemic insecticide/nematicide registered on many ornamentals for controlling a broad spectrum of insects and mites. Oxamyl may be incorporated into media before planting or broadcast on the media after planting. Vydate may be used as a spray, soil drench, or root, bulb or corm dip.

Oxythioquinox (Joust, Morestan), class: dithiocarbonate. These products are registered for control of whitefly eggs and other immature stages on outdoor crops only. Use at the lowest rate possible because of possible phytotoxicity.

Permethrin (Pounce, Pramex), class: pyrethroid. Pramex has a broad registration on ornamental plants for controlling a range of pests, including whiteflies and caterpillars. Pounce and Pramex are registered for leafminer control on greenhouse chrysanthemums. Pounce is also registered on field- and greenhouse-grown roses for controlling several Lepidoptera larvae.

Pyrethrins (PT 1100, Pyrenone Crop Spray, Pyreth-it, X-Clude), class: botanical. Pyrethrins are available in several formulations that may or may not be synergized with piperonyl butoxide (PBO). Pyrethrins are especially effective against small, flying insects. X-Clude is a microencapsulated formulation, available as a liquid concentrate and aerosol.

Resmethrin (PT 1200, SBP 1382), class: pyrethroid. Resmethrin is an early generation pyrethroid insecticide registered for use as a foliar spray or aerosol. It's effective against pests such as whiteflies, although some resistance has been reported. Best results occur with cool greenhouse temperatures (50 to 75 F).

Sulfotepp (Dithio), class: organophosphate. Sulfotepp is a vapor-active pesticide registered for use as a smoke generator. It has been one of the most effective pesticides to control sweet potato whitefly on poinsettia and other crops. At present, it's registration status is in doubt.

Chapter 18

Integrated Pest Management

During the past few years, greenhouse crop producers have had to contend with a number of "new" insect and mite pest problems. These pests include leafminers, western flower thrips and sweet potato whiteflies. These pests, plus the other pests that normally are present, have put a real strain on existing pesticides and other pest management methods, stimulating interest in future prospects. This chapter looks ahead to the next decade on the subjects of insect and mite pests, pesticides and alternate controls.

Insect and mite pest problems

Because of the tremendous worldwide transport of plant materials, pests will continue to appear very suddenly in areas far from their native habitat and cause severe injury before control programs can deal with the problem. Some pests will escape detection by plant inspectors, and even the best quarantine systems can't prevent pest movement—they can only slow it down. Thus, there will be "exotic" pests to contend with.

Discounting actual or potential problems with imported exotic pests, the insect pest groups causing the most severe economic losses in the near future probably will be aphids, thrips and whiteflies. Problems with these three groups will be caused by pesticide resistance, both behavioral (avoiding contact with pesticides) and physiological (insect or mite populations breaking down the pesticide chemically). The green peach aphid, greenhouse and sweet potato whiteflies and western flower thrips have all developed resistance to several pesticides in different chemical groups.

Insecticides and miticides

The registration of new, conventional insecticides and miticides has slowed considerably in recent years. All pesticides presently registered in the United States must be re-registered by the late 1990s, and several registrations on ornamental plants will be dropped. This will happen simply because the costs

Insects caught on sticky trap.

of obtaining the additional data to meet new requirements is too great for the ornamental plant market size. New registrations have occurred among the advanced pyrethroid pesticides such as fenpropathrin (Tame) and cyfluthrin (Decathlon, Tempo). Most registrations are in other pesticide classes, however. Insecticidal soaps and horticultural oils have reappeared on the market recently. Several insect growth regulators, including cyromazine (Citation), and kinoprene (Enstar) are registered and available for use. Another pesticide group consists of natural products, such as the insecticide/miticide abamectin (Avid), which is derived from toxins produced by soil microorganisms and acts on the insect or mite differently from more conventional materials. Another natural product is based on extracts of seeds from the neem, or margosa, tree. Many other insecticidal plant extracts exist and await commercial development.

Other developments are in the microbial pesticide group. *Bacillus thuringiensis* (Dipel) is a charter member of this group, but others include *Bacillus thuringiensis* H-14 (Gnatrol). Gnatrol controls pests such as fungus gnat larvae. More such "biological" or "natural" pesticides may be developed, including fungi. Commercial formulations of *Verticillium lecanii*, which attack certain aphid, whitefly and thrips species, are being sold in Europe. Other fungi include Aschersonia, which attacks greenhouse whiteflies, and *Paecilomyces fumosoroseus*, which attacks sweet potato whiteflies. Development of products will proceed if potential markets exist and the registration process is able to deal with the nonconventional products. Sometimes the idea is wonderful but the potential profit and/or patent protection isn't there.

The best thing about these newer pesticides is that they tend to be "softer" on the environment, including beneficial insects and mites, than many conventional materials. They can be used in combination with biological control programs that include predators and parasites. Although these materials are effective, users need education about proper application and expected results.

Even if pesticide registration were continuing at a normal pace, many new regulations have been, and continue to be, implemented by national, state and

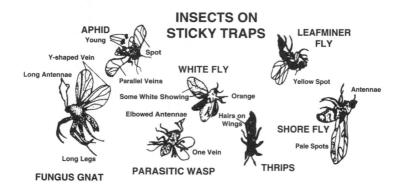

Outline of several pest types on sticky traps.
Courtesy of James R. Baker, North Carolina State University.

local governments in areas of worker safety and right to know. It's difficult to argue *against* safety, so we have to accept the fact that rules and regulations are necessary to protect us from ourselves. Common sense would dictate safe handling and use (pesticides are designed to kill things), but not all of us seem to be blessed with it. Thus, more laws are passed. In most cases, re-entry and/or safety regulations don't cause major disruptions in normal operations. Another important aspect of the pesticide situation is the public perception of pesticides. Thus far, this hasn't extended to pesticide use on ornamental plants, but there's no predicting where this movement will stop. All of the above situations have led toward investigating other ways to deal with insect and mite pests. These methods can be grouped under the heading of integrated pest management.

Integrated management of insects and mites

Integrated pest management means using as many methods as possible to minimize insect and mite problems. These methods can include trapping and scouting for pest detection, as well as cultural, physical, environmental, chemical (pesticide) and biological controls. These components, discussed individually below, must be used in a *systematic* way. Planning and record-keeping are required. One of the first things to do when implementing an IPM program is to eliminate the concept of zero pests. We cannot achieve this even when overusing pesticides, and the goal of most IPM systems is to use fewer pesticides (although pest levels should not be markedly higher than before). Another concept to change is the attitude that IPM is somehow more "risky" than strict reliance on pesticides. Certainly, jumping directly into biological control from a pesticide control program without adequate preparation and knowledge will likely cause major problems. These items can be classified into the grower attitude category. A proper attitude on the part of growers and managers is necessary if IPM is to succeed. Following is a discussion of integrated pest management of insect and mites.

Yellow sticky trap.

Pest detection

The most critical part of any pest management program (whether IPM or reliance only on pesticides) is knowing what pests to expect and/or detecting pests at a very early stage. Inspecting plants regularly and/or using traps for flying insects allows early detection.

During the past few years a great deal of discussion has concerned using yellow, sticky traps to monitor and/or reduce certain insect pest populations. Yellow traps catch winged aphids, leafminers, thrips, whiteflies, fungus gnats and shore flies (among others). Blue traps are sometimes better at detecting western flower thrips, but yellow is satisfactory. Once trapped, the pest insects must be identified. Recognizing these pests may take some training and a 10X or 15X magnification hand lens, but it's very important to learn the general outlines and colors of different pest groups.

The best reason to use sticky traps is to track insect population trends (those mentioned above) in your greenhouse. Trap catch records provide information on when to apply pesticides and let you know how the control program is going. The traps won't totally replace plant inspection, but, if used correctly, can be valuable additions to a pest scouting program. Traps are generally not used to reduce an insect population, but as a scouting tool. Examine traps weekly. Many growers/managers will collect old traps and make actual counts of insects for future reference. Others make estimates. Traps can then be cleaned or replaced (see below). Use some system of numbering and greenhouse mapping to keep records of trap locations and insect infestations. This record-keeping should be a major part of any pest management program. At first the records of trap catches may not be meaningful, but keep in mind that they are part of an organized system. Over time, these records will be very useful.

Place traps vertically at or just above plant height, the location of most "action" in insect flight activity. Fungus gnats, shore flies, thrips, sweet potato whiteflies and leafminers can also be trapped just above the growing medium surface on horizontal traps. Be sure to place some traps near side vents, doors and known susceptible cultivars. The number of traps necessary will depend

Unharvested crop, a source of insects and mites.

upon your objectives and ability to inspect them. A minimum number is four to six per acre, but more will be better, especially if using traps to monitor whitefly population trends.

Within limits, trap size isn't that important. Smaller traps are more efficient, perhaps because there is more edge than surface area (greater numbers of insects trapped per square inch on smaller traps, compared with larger traps), but large traps are also effective. They can be square, rectangular or cylindrical. You can purchase traps from many greenhouse equipment supply outlets or make them in-house. Effective traps can be made out of almost anything that either is or can be painted bright yellow. Rust-Oleum No. 659 Yellow is a bright yellow paint that works quite well. The sticky material can be as simple as a thin layer of cooking oil, other oil, a mixture of mineral oil and petroleum jelly, or commercially prepared materials such as "Sticky Stuff" (Olson Products Inc., P.O. Box 1043, Medina OH 44256) or "Pestick" (Phytotronics Inc., 2760 Chouteau Ave., St. Louis, MO 63103). Try placing the sticky material on clear plastic wrap and covering the yellow surface with the wrap for easier removal and trap changing. Remember that sticky traps won't replace plant inspection in most cases; use the two together. Whiteflies tend to occur in localized infestations that traps may not detect. Non-winged aphids and spider mites aren't caught on traps. To summarize, the steps are:

A. Use yellow traps as a pest management tool.
B. Learn to identify what you catch.
C. Check, record and/or change traps weekly.

Remember that using traps won't replace plant inspection, but are effective when used along with it. Inspect plants in every bed or on every bench weekly. Inspect plants along outside walls and near open vents and entryways most thoroughly. Also, inspect any new plant shipments closely before placing them in the main production area. Mark areas found to be infested with a particular pest (use small flags or colored ribbons). Some growers use a different color to indicate each pest type found. This will enable you to do follow-up inspections after applying pesticides or introducing natural enemies. As with trapping, the number of plants to inspect varies with the greenhouse operation and the inspector. Several states have organized IPM programs for greenhouse crops. More will be implemented soon. These programs include regular visits by trained pest management scouts who do all the above trapping, plant inspections and record-keeping. Private pest management companies may also have the capability to do this in some areas. There is a cost for these services, but experience has shown that IPM programs are cost-effective. It's unlikely that every ornamental plant producer will be part of an organized IPM program; perhaps IPM won't be practical for everyone. But if no pest management service is available, the grower may decide to assign one or more employees to the task. Either way, it will be quite useful in many production operations.

Cultural/physical control

These methods are very important. Used alone, cultural or physical controls probably won't do the entire job of pest management, but often will be effective in reducing pest numbers to allow other methods to be more effective. Start with clean plants from reputable suppliers and inspect them thoroughly. Avoid buying plants from other growers to fill orders unless you have inspected them at the other grower's establishment. If you've opened boxes inside your greenhouse and pests have flown out or you've put infested plants in the greenhouse, it's too late.

Select cultivars or plant species carefully. If western flower thrips are a severe problem for you, avoid producing crops that are extremely susceptible to thrips attack and/or virus expression. If possible, produce pest-resistant cultivars. Unfortunately, few cultivars with resistance to insects and mites are available, but this aspect of pest reduction is receiving more attention. For example, certain schefflera cultivars resist two-spotted spider mite attack. Spider mites are among this plant's most important pests. Surveys of important ornamental plant species cultivars to determine pest preference have sometimes indicated large differences in pest populations. Removing the most preferred cultivar from production won't necessarily lower the pest level. The pest may simply move to the next cultivar on the list. But knowing something about the pest susceptibility of cultivars that you are producing will help you make educated guesses concerning where pests will be more serious. Try to group the most susceptible cultivars in areas physically separated from other crops or cultivars or at least in one zone within the main production area.

Destroy crop residues promptly after harvest. The longer these plants remain, the greater the chances for increased pest populations. For example, during warm weather the western flower thrips can complete a generation in about one week. Don't place crop residues outside the greenhouse. The flying adults from many insect species will simply move back into the greenhouse. Don't plant a vegetable garden adjacent to the greenhouse, for the same reasons.

A period of weeks without any crop production, providing that all weeds are removed from inside the greenhouse, would be of tremendous help in "cleaning" up the greenhouse, disrupting pest populations and allowing the subsequent crop to be virtually pest-free at the beginning. But for economic reasons, growers want to keep the greenhouse full of plants. This may be good for projecting potential income per square foot and, in the absence of severe pest problems, probably is fine. If pest carryover between crops becomes a problem, however, a crop-free period will be necessary.

Use screens or other barriers to restrict pest movement into the greenhouse from outside or between different production areas within the greenhouse. Several types of screening material are available, some with mesh size small enough to restrict thrips movement. This naturally will restrict air movement as well, but can be compensated for by placing the screening on a frame or "cage" over the air intake to increase the surface area through which the air is drawn. Greenhouse supply companies that sell these materials will be able to assist individual growers who wish to use screens. See Appendix I for a list of some manufacturers.

The surface on which plants are produced may be modified to discourage pest development. Several examples involve plastic under plants or pots. Again, this approach isn't new, but with increasing numbers of growers producing plants in materials such as rockwool on plastic sheeting, opportunities are presented that may help to control pests such as western flower thrips and leafminers. A possible pest management method here is to use water. For example, leafminers and thrips generally drop off plants and pupate (transform to adults) on or in the medium below the plants. On plastic, the larvae seek the lowest point. Thus, periodic wetting of the plastic substrate will drown significant numbers of leafminers and thrips. For growers using ebb and flood irrigation systems, many pests might be washed away along with the irrigation water. Of course, many pests also drop into pots or remain on leaves and flowers.

Environmental control

Manipulating the greenhouse environment has long been suggested as a way of preventing establishment of plant pathogens; for example, venting and heating at sundown to reduce leaf wetness helps prevent fungal spore germination. Some commercial greenhouse rose production programs use leaf wetness sensors connected to a greenhouse environmental control computer system to prevent powdery mildew spore germination. This technology will be more useful as more environmental control computer systems are installed in commercial greenhouses. To date, very little work has been done on manipulating the greenhouse environment to make it less favorable for pest insects and mites and/or more favorable for natural enemies, even though certain environmental conditions favor establishment of insect-pathogenic fungi, and establishment and dispersal of introduced natural enemies. For example, thrips and spider mite numbers tend to be lower on crops grown under a regular daytime misting (to wet the leaves) regimen. Spider mite predators can better control the mite population under these conditions. The difficulty here, of course, is to integrate the conditions best for the insect/mite management system with those that don't promote establishment of plant pathogens. It won't be possible to do this on all crops, but it may be possible on some.

Biological control

There is increasing interest in using biological control agents on greenhouse ornamental crops in the United States. Some parts of the world (Northern

Europe, Soviet Union, Canada) have well-established programs for biological control of major insect and mite pests. Nearly all successful biological control programs are on greenhouse vegetable crops rather than ornamentals. The same pests usually occur on ornamentals, but vegetable crops usually have a higher tolerance for insect or mite presence and can usually tolerate some feeding injury and still produce high yields. Even though most ornamental plants also tolerate relatively high levels of many pests without showing visible injury, most producers and consumers aren't willing to tolerate these levels for aesthetic reasons. Despite such difficulties, some ornamentals are legitimate candidates for an insect and mite biological control program. These include cut flower crops such as chrysanthemum, gerbera and rose, as well as pot-produced crops such as poinsettia.

Biological control agents in this chapter are defined as using living organisms to control or manage an insect or mite pest. Included are parasites, predators, fungi, nematodes and viruses. Many of the most important biological controls available are listed in the discussion of individual pests in Chapter 16. The biological control agents generally must be purchased from specialty producers (see Appendix I), and released onto the crop regularly. They are used in a program similar to a regular insecticide application program. Even more than with pesticides, pest management will fail if pest populations are too high when the biological control agents are introduced. Biological controls aren't rescue treatments. Often, pest populations won't decline, and probably will even increase, for several weeks to two months after introducing natural enemies. First, keep in mind that, as with IPM in general, biological control requires a different outlook on pest management, and the speed of pest control. Important factors include planning and pesticide use. Biological control, although *theoretically* possible for most pests on most crops, won't be practical in some situations. As desirable as the concept is, reality needs to enter our thinking. Answer the following questions for your greenhouse operation to determine the feasibility of biological control:

A. How many crops are produced? What crops? What is your tolerance for crop damage? What about your customers' tolerance?

More crops often mean a more complicated pest situation. Some crops have an extremely low tolerance for pest injury. The crop and how it's produced is very important. Pot chrysanthemums generally will have less tolerance for injury to lower leaves by leafminers compared with chrysanthemums produced for cut flowers. The lower leaves on cut flower chrysanthemums can be removed. Gerbera and rose (produced for cut flowers) are other candidates. One of the few pot-produced crops where biological insect control may be practical is poinsettia. Few or no economic injury levels are established for most ornamental plants. In the case of western flower thrips and spotted wilt virus, the pest level tolerated is as close to zero as can be obtained. Some people have a low tolerance. For example, some growers are not willing to tolerate even one adult whitefly.

B. Are individual crops produced in separate areas or in one large greenhouse without physical separation?

The most difficult situation for biological control is a greenhouse that produces many crops continuously without any walls or other barriers to separate them. The lack of separation makes it impossible to control both pest and natural enemy movement within the greenhouse. If screens or other barriers cannot be erected, biological control will likely fail. Pesticide applications in other greenhouse areas are often harmful to natural enemies, even without direct contact with the spray. See the following pesticide discussion and table for information on the effects of certain pesticides on selected natural enemies.

C. What are the major insect and mite pest problems?

This is, of course, governed by the crops produced and geographical area of the greenhouse, but biological control is usually most successful on crops attacked by only one or two major pests. Even though natural enemies are available for nearly all major pests, the difficulty of manipulating this complicated system will be too great for most people.

If the crops produced and production system seem satisfactory for a biological control program, contact a reputable supplier of natural enemies several months before beginning a biological control program. Contact your state extension entomologist for this information, or see the list of these suppliers in Appendix I. The supplier should know what crop(s) you are producing, likely pest (and disease) problems and approximate numbers of natural enemies that will be required. You will probably need to seek the advice of a knowledgeable expert (the natural enemy supply company, an experienced grower or consultant). Try to get advice from someone with experience in your growing situation and/or general geographic area. Valid advice in the Netherlands may not be relevant to California, Florida or northern states. Several states now have organized programs to implement biological or integrated pest management systems on greenhouse crops. Advice during the crop production period is also very important. Have some idea how long it should take to see results. Remember that several weeks to months will be required to lower an insect or mite pest population with any natural enemy.

Use biological controls on a small area at first, expanding later as circumstances permit. If your production areas aren't set up for this, construct a special "biocontrol" house or area for this purpose.

Some commercially available biological controls

Common name	Scientific name	Pests attacked
Aphid midge	*Aphidoletes aphidimyza*	Aphids
Green lacewings	*Chrysoperla* sp.	Aphids, caterpillars, mites, whiteflies
Leafminer parasites	*Dacnusa* spp., *Diglyphus* spp.	Leafminers
Mealybug destroyer	*Cryptolaemus montrouzieri*	Citrus mealybugs
Minute pirate bug	*Orius* spp.	Thrips, other pests
Nematodes	*Steinernematidae*	Beetle larvae, fungus gnat larvae
Spider mite predator	*Phytoseiulus persimilis*, other phytoseiids	Spider mites
Thrips predator	*Amblyseius cucumeris, A. barkeri*	Thrips
Whitefly parasite	*Encarsia formosa*	Whiteflies, especially the greenhouse whitefly

Pesticides

Realistically, biological control can't generally be used for one pest, and pesticides to control others. Most insecticides are harmful to parasites and predators (see the following table). Some have effects for weeks after application. Be *extremely careful* when using these materials. Don't use them for at least 30 days before beginning a biological control program, and don't use them anywhere in the greenhouse after beginning a program. Avoid using materials such as methomyl, malathion and any pyrethroid, as these tend to be most harmful. Experts differ in determining how harmful specific pesticides are, but it's always best to err on the conservative side. If possible, don't apply insecticides to the entire crop area. Local or spot applications generally are less harmful to natural enemies. This will negate the usefulness of many of the low-volume methods described in Chapter 22, which are designed to apply pesticides over large areas.

Some pesticides are less harmful. Many fungicides can be safely used, but some are harmful. As mentioned at the beginning of this chapter, many of the newly-registered pesticides are "soft" on beneficial insects and mites. *Bacillus thuringiensis* formulations are harmless to natural enemies. Insecticidal soaps, as well as insecticides such as abamectin, neem extract, cyromazine and kinoprene, tend to be less harmful than conventional materials. This doesn't mean that these materials are harmless, but that natural enemies can be reintroduced after an application without any negative effects.

Effects of some insecticides and fungicides
on biological control agent

Pesticide trade name	Generic or common name	*Encarsia formosa**	*Phytoseiulus persimilis**
Ambush, Pounce, Pramex	Permethrin	H (>30)	H (30)
Botran	Dicloran	S	S
Bravo, Daconil	Chlorothalonil	S	S
Captan	Captan	S	H (7)
Decis	Deltamethrin	H (>30)	H (>30)
DDVP, Vapona	Dichlorvos	H (3)	H (3)
Diazinon	Diazinon	H (14)	S
Dibrom	Naled	H (7)	H (7)
Dipel, Victory	*Bacillus thuringiensis*	S	S
Dithio	Sulfotepp	H	H
Lindane	Lindane	H (>30)	H (7)
Malathion	Malathion	H (14)	H (14)
Manzate 200	Manzoceb	S	H
Pyrenone	Pyrethrum + PBO	H (10)	H (10)
Ridomil, Subdue	Metalaxyl	S	S
Rovral	Iprodione	H (0)	H (0)
Sevin	Carbaryl	H (30)	H (30)
Sulfur, Flotox	Sulfur	S	S
Sunspray oil	Petroleum oil	H (0)	H (0)
Thiodan	Endosulfan	H (unknown)	H (unknown)
Vendex	Fenbutatin oxide	S	S

*H = harmful; S = safe; ( ) = number of days after application that a material remains harmful. Some materials, such as sulfur sprays and dusts, aren't directly toxic, but, especially after repeated use, will interfere with biological control programs.

Application method and formulation will affect the toxicity to natural enemies. A number of studies have conflicting information concerning how long a particular pesticide will remain harmful to a natural enemy. When in doubt, err on the conservative side.

Table modified from information supplied by Applied Bio-Nomics Ltd., Sidney, B.C. Canada.

Chapter 19

Common and Trade Names of Insecticides and Fungicides

*T*he names of pesticides used on greenhouse crops can be confusing because of the combination of common names and various trade names. Literature referring to the usefulness or hazards associated with one or another product will often refer to one or another of these names. European trade journals can be a particularly useful source of such information. In addition, if you receive plants from European sources or from different parts of the United States, you may be concerned about what has been previously used on the plant material. The following tables may help you to discern the pesticides used. Of course, not all trade names are included in these tables. Most of the U.S. trade names have been listed in connection with disease and insect management in previous chapters, so the European trade names will be of greatest interest. Only the trade names used most often are included. The names constantly change and come and go from the marketplace. The following tables summarize current information.

Insecticides and miticides

Common name	Trade names in United States	Trade names in Europe
Abamectin	Avid	Vertimec
Acephate	Orthene, PT 1300	Tornado
Bacillus thuringiensis	Dipel, Sok-B.t., Thuricide	Tribactur
Bendiocarb	Dycarb, Ficam, Turcam	Garvox
Carbaryl	Sevin	Carbaryl
Cyfluthrin	Decathlon, Tempo	Baythroid
Deltamethrin	Decis	*
Diazinon	Diazinon, Knox-Out	Basudin, Basudine, Diazin
Dichlorvos	Vapona	Nogos, Nuvan

Common name	Trade names in United States	Trade names in Europe
Dicofol	Kelthane	Acarfen, Akatox
Dienochlor	Pentac	Dienox
Dimethoate	Cygon	Perfeckthion, Rogor
Disulfoton	Di-Syston	Solvirex
d-Phenothrin	Sumithrin	*
Endosulfan	Thiodan, Tiovel	Afidan, Endamon, Cloaca
Fenbutatin-oxide	Vendex	Torque
Fluvalinate	Mavrik	Klartan
Kinoprene	Enstar	*
Malathion	Cythion, Malathion	*
Metaldehyde	Deadline, Slugit, Snarol	Antimilace, Cekumeta
Methiocarb	Grandslam, Mesurol, PT 1700	Draza
Methomyl	Lannate	*
Naled	Dibrom	Bromotox, Pyomix
Nicotine sulfate	Nicotine, Black Leaf 40	Hypnol, Nikotoxin
Oxamyl	Vydate L, Oxamyl	*
Oxythioquinox	Joust, Morestan	*
Permethrin	Ambush, Pounce, Pramex	Coopex, Kafil, Permasect, Perthrine, Talcord
Pirimicarb	Pirimor	Aphox
Pyrethrins	PT 1100, Pyrenone Crop Spray, Pyreth-it	Fogox, Pybuthrin, Pyrel, Pyrsol, Zyrpytrine
Resmethrin	PT 1200, SBP 1382	Chryson
Sulfotepp	Dithio	Bladafum

---

*Widely listed under same trade name(s) as in United States.

Fungicides and bactericides

Common name	Trade names in United States	Trade names in Europe and elsewhere
Bordeaux mixture	Bordeaux Mix, Bordocop, Bordo-Mix	Comac, Nutra-Spray
Captan	Captan, Orthrocide	Captab, Captane, Captanex, Merpan, Vondcaptan
Chlorothalonil	Bravo, Daconil 2787, Exotherm Termil	ClorotoCaffaro
Cupric hydroxide	Blue Shield, Champion	ComacParasol, Criscobre, Cudrox, Cuproxide, Manpower Hydrocop
Dicloran	Botran	Allisan, Resisan
Dinocap	Karathane	Crothothane
Dodemorph	Milban	Meltatox
Etridiazole	Koban, Truban	Aaterra, Dansoil, Pansoil, Phorate TSX
Ferbam	Carbamate	Ferberk, Hexaferf, Knockmate Trifungol
Folpet	Phaltan	Folpan, Thiophal
Fosetyl-Al	Aliette	*
Iprodione	Chipco 26019, Rovral	Glycophene
Mancozeb	Dithane M-45, F-45	Cuprocas, Frugold, Galben M, Mancofol, Mancogin, Manzate, Nemispor, Nespor, Penncozeb, Policar, Poliman, Tairel M, Vandozeb Plus
Metalaxyl	Apron, Ridomil, Subdue	*
Oxycarboxin	Plantvax	Oxykisvax
Piperalin	Pipron	*
Propamocarb-HC1	Banol	Dynone N, Filex, Prevex, Previcur
Propiconazole	Banner, Tilt	Desmel
Quinomethionate	Morestan	Bay 36205, Chinomethionate, Quinomethionate
Quintozene	PCNB, Terraclor	Avicol, Botrilex, Earthcide, Folosan, Kobutol, Pentagen, Tilcarex, Tri-PCNB
Streptomycin	Agri-Mycin 17, Agri-Step	Plantomycin

Sulfur	Brimstone, Flotox, Sul-Cide	Cosan, Ditiozol, Elosal, Frugold, Kumulan, Sulfex, Thiolux, Thion, Thiovit, Zolfo
Thiabendazole	Arbotect, Mertect	Tecto, Thibenzole Starite, Storite
Thiophanate-M	Cleary's 3336, Domain, Domain FL, Systac 1998, Topsin M	Cercobin-M, Homai, Labilite, Mergibon, Mildothane, Sip-casan, Thiophan
Thiophanate-M plus mancozeb	Duosan, Zyban	*
Thiophanate-M plus etridiazol	Banrot	*
Triadimefon	Bayleton	Amiral, BayMeb 6447
Triforine	Triforine, Funginex	Cela W-524, De-narin, Saprol
Vinclozolin	Ornalin, Ronilan, Vorlan	*
Zineb	ChemZineb, Parzate, Polyram Z, Zineb	Aspor, Bleu, Carbina TZ, Cumene, Cupsin 60, Diphen, Ditiozin F80, Ditiozol, Enoz in 80, Galbenz, Kypzin, Lonacol, Mexathane, Tanazone, Tarelz, Tiazin, Tiezene, Tritoftorol, Zinecu-pryl, Zinosan

*Widely listed under same trade name(s) as in United States.

Chapter 20

Pesticide Labels, Safety and Common-Sense Chemistry

*P*esticides, because they are products offered for sale for a certain purpose, have labels that serve various functions. These labels advertise and make the product seem attractive and beneficial. They also teach you how to use the pesticide safely and effectively. This second label function is fairly standard from product to product because it must meet the approval and follow the guidelines set up by the Environmental Protection Agency. Using a product efficiently, safely and effectively depends on your familiarity with the label.

How to read a pesticide label

Pesticide labels are divided into several sections. The following outline will briefly describe each section. There are many different styles and forms of the various sections and the order of the sections may not follow the order used here.

Trade name and basic function

Generally the most prominent part of the label is the product's trade or brand name. The manufacturer wants you, as a user, to become familiar with the product as he is selling it. The trade name may indicate the manufacturer. The manufacturer's address must also appear on the label, but this is often separate from the trade name section. Some statement about the product's basic function is generally close to the trade name. For instance, you might see the term, "an emulsifiable concentrate insecticide for use on house plants," but you won't precisely know the product's function from this initial statement.

Ingredients and formulation

On the package front is an ingredients statement, usually right under or quite close to the trade name. Each ingredient and its percentage of the product is listed. The percentage must come to 100 percent, but the companies are required to list only the ingredients that act as pesticides. All other ingredients, such as stabilizers or wetters, are listed together as "inert ingredients." The active ingredient is identified by either a long, rather complicated chemical name or as a shortened name, the common name of the pesticide. Sometimes both names will be listed, but not together. Look for asterisks or footnotes to provide more ingredient information on the label.

The ingredients statement can sometimes be helpful because it can tell you how one product may compare with another. Are they in fact the same product but under different brand names? Do they differ in strength? Do they contain the same kind of pesticide? Are they combination products with a certain familiar pesticide combined with some that aren't familiar? It's good to understand the ingredients statement and be able to sort through it because you might be surprised how much you already know about a seemingly new or unfamiliar product.

Near the ingredients statement appears the EPA establishment and registration number. These official numbers appear on government documents relating to the product. If there are serious questions about health hazards or plant damage resulting from using the product, consult with experts by referring to this registration number.

Health and safety information

The front panel and part of the side panels of an insecticide package contain health and safety information. The statement "keep out of reach of children" is required on all pesticide containers. Other general statements provide the chemical's toxicity hazard category. The words "danger," "poison," "warning" or "caution" may appear. Any special hazards are prominently displayed in this health and safety information section. For instance, if the product is a restricted-use pesticide available to licensed applicators only, this information will appear quite prominently on the front of the package. If it's a specifically dangerous chemical (causes eye damage, skin rash or breathing difficulties), this information appears prominently in the health and safety section. Antidotes or treatments in case of accidental exposure are often given. Often, this health and safety information will be boxed in a red or black outlined section of the label.

Directions for use

This section can be quite detailed and lengthy, containing several parts you should look for. First various plant types are listed. Sometimes it's confusing when a big category is listed and then specific plants follow behind in parentheses. Actually, when such a listing is made, the product is registered or safe for only specific plants and not the big category. A statement may read, for example, "foliage plants (ficus, schefflera, pothos)." This is somewhat confusing and leads to a real question as to what precisely is labelled! The safest conclusion is that unless the plant is specifically mentioned on the label, it's not necessarily included.

The next section of the directions for use is the list of pests or diseases that can be managed or controlled with the product. These may be listed specifically by each plant, or they may be listed in a separate section. Sometimes the scientific Latin names will be given for individual pest species or genera. Other times, just a general type (such as aphids) will be listed.

Dosages and dilutions for spray preparations will be given in this label section, followed very closely by application method. High volume, spray-to-runoff directions are given most commonly for products used on ornamental plants. Sometimes you will also see special directions for applying low-volume, spray concentrates. You can learn a lot about a product's general usefulness by carefully reading the directions-for-use section of the label.

Phytotoxicity or chemical mixing problems

This portion of the label is quite closely related to and sometimes included in the directions-for-use section. This very important section includes possible problems you might encounter with using a particular pesticide. It may warn you not to use the product on certain plants or caution you about using the product with another pesticide in a "tank mix." It may give you guidance about which products you can successfully mix.

As you go through some example labels, you begin to become familiar with their general layout. Knowledge of the label is critical to effectively use the chemical. A good review of a label can be very important before you buy the product. It may enable you to make some decisions about how useful the product can be in your pesticide program for the plant material you are treating.

Understanding pesticide safety

Pesticide safety can be separated into three areas. The first area is acute toxicity. The second area encompasses chronic toxicity. The third area involves hazards of use. Each area has more or less importance to us in the field or greenhouse.

Acute toxicity

As the name implies, acute toxicity is the ability of the pesticide to cause an acute, toxic reaction immediately or soon after exposure. Such reactions include shortness of breath, tunnel vision, skin problems or eye, ear, nose or throat irritation. Acute toxicity reactions are often confused with allergic reactions. An allergic reaction is one that may occur in some people after an initial sensitization exposure. It tends to be independent of exposure dosage and might occur after even minute exposure! Acute toxicity reactions are always related to dosage, although some people are more sensitive than others.

Acute toxicity is usually measured by a figure that's called LD 50. LD 50 is a population toxicity measurement. It's the dosage level of a particular toxin that will kill 50 percent of a test population of animals. LD 50 is measured by a dosage in relation to the test animal's weight. It's expressed as milligrams per kilogram of body weight. It takes more of a particular toxin to kill a larger organism or a bigger human being than it does a smaller one. This is because of the body's ability to dilute the toxin as it enters the body's biochemical pathways. Dilution makes the dose less harmful; however, since LD 50 is the fraction or ratio of toxin over weight, it's a useful number regardless of the size of the person using the product. It can be measured by different kinds of exposure paths, such as dermal exposure, inhalation exposure or ingestion exposure.

LD 50 is a useful device for a chemical company because it tells the company in general how toxic their chemical will be. They, of course, are quite interested in general toxicity because they are selling the product to a "population" of users. If one chemical is generally more toxic than another, they are interested in labeling and packaging it more carefully. For instance, the word "warning" and a skull and crossbones on the label means that the pesticide is a category 1 pesticide of relatively low LD 50. Generally, those chemicals with lower LD 50s are more dangerous.

Acute toxicity information is important to us in a general sense. Many chemicals that we come in contact with can be acutely toxic to us if used carelessly or stored improperly. In many cases their toxicity level exceeds that of pesticides you may be using. The table below lists some of these common chemicals. The table is presented not to imply that pesticides are safe but to illustrate that pesticides can be no more or less toxic to you than other chemicals in our world.

Acute toxicity levels of common chemicals

Material	Oral LD(50)*
Systox	6
Nicotine	10
Kerosene	50
Metasystox	65
Rotenone	75
Diazinon	108
Gasoline	150
Caffeine	200
Sevin	850
Kelthane	1100
Aspirin	1200
Malathion	1375
Pyrethrin	1500
Table salt	3320
Maneb	6750
80 Proof alcohol	13300
Captan	15000

*Measured in milligrams per kilogram of body weight.

In the above table, it generally takes more of the products with larger numbers to hurt us than it does for the products with lower numbers. We might be inclined to think of those chemicals with lower numbers as more dangerous; however, this has little meaning to us on an individual basis. Any individual might be quite susceptible to the toxic property of a pesticide even though it may have a rather high LD 50. For this reason, LD 50s aren't placed on pesticide package labels. You should be careful with any pesticide.

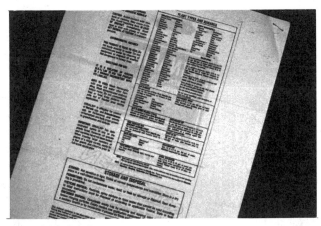

A pesticide label.

Chronic toxicity

The second area of pesticide safety is chronic toxicity. Chronic toxicities are those having long term effects. They largely result from long term exposure. Such exposure may be in minute dosages, but if it accumulates in an organism over its lifetime, it results in chronic toxicity problems. Chronic toxicities include cancers, birth defects or reduced fertility levels.

Two inherent problems of chronic toxicity make it difficult for scientists to reach sound conclusions about how it relates to us as users. First, proving that a substance does *not* cause a chronic toxicity problem becomes a very difficult exercise in proving a negative. Actually, it's impossible for the scientific method to prove a negative. As data masses become large enough, most people will accept the "fact" that something doesn't occur. For example, have we actually *proven* that spontaneous generation does *not* occur or that Santa Claus is *not* real?

The other problem with chronic health effects and how they relate to you as a pesticide user centers on the dosage effect. Many compounds are tested for chronic toxicity problems at the highest tolerable dose of the test population. This is a dosage much higher than any dosage anyone would ever be exposed to in use once the materials are registered, labelled and released into the market-place. If a problem occurs at the massive dose, then we must conjecture about what the result would be at a very minute dose. Many people believe that even a minute dose will have a small, but noticeable effect on the population in general. Others believe that as the dosage goes down, the chemical reaches a zero effect or threshold level. The point is, it's practically impossible to define these curves, because we have no means of gathering data on minute dosages in human populations. It's especially troublesome when we don't really know all the causes for chronic disorders such as cancers or birth defects. The issue remains resolved in scientific circles!

The decision whether to use a product implicated in chronic health effects at massive dosages needs to be made by individuals with proper understanding of the problems. Generally, chances of contracting chronic problems even in the test animal populations are extremely minute, much smaller than any we would normally encounter with many things that we do in every day life. For instance, it has been recently calculated that a person applying 2,4,5-T herbicide five days a week with a back-pack sprayer, four months a year for 30 years would have a 0.4 per million chance of developing a tumor. This can be compared with smoking cigarettes (1,200 chances per million), being in a room with a smoker (10 chances per million), drinking one can of beer per day for 30 years (10 chances per million) or sunbathing (5,000 chances per million). One can easily recognize that these statistics don't necessarily help analyze personal risk. Unfortunately, science cannot adequately address this issue!

Hazards of use

The third area of pesticide safety that relates to those in the field or greenhouse is hazards of use. Reducing hazards of use is the most important thing you can do to ensure and improve pesticide safety. Hazards of use are primarily those involving exposure problems. How much pesticide is getting into a worker's system because of ways of handling, storing, preparing, using and disposing pesticide preparations? It's a good general practice to remind yourself to always try to reduce pesticide exposure. Take this point of view regardless of how inherently toxic the product is in the two areas already mentioned.

Recent research on exposure hazards to pesticides has had interesting results. In most of these research efforts, scientists monitor people during the normal patterns of using the product and measure the pesticide that is picked up on workers' hands, face, clothing and lungs. Several common denominators from most of the studies give us keys to reduce hazards of use. An outer layer of clothing worn over regular clothing when applying pesticides is a primary means of reducing exposure hazards. This outer layer can be a special spray suit or simply pants and shirt that can be removed and laundered after each pesticide exposure. Follow labeled instructions when given.

Many use and exposure studies done around the country show that much pesticide exposure occurs during mixing and preparation. It's easy to get high pesticide dosages into your system when you're mixing and measuring concentrates as they come from the producer. Many agricultural industries have adopted closed mixing technologies where no pesticide comes in contact with the worker or the air. The systems may include water soluble packets for wettable powders or automatic can and drum openers that work by remote control and pump the liquid through closed tubing directly into the spray tank. In the future we will probably see more of these systems in all areas of agriculture. Hands and forearms are often implicated as sites of large exposure during these mixing and preparation operations. Thus, you can see the tremendous value of wearing gloves and long sleeved shirts or coveralls when handling pesticides.

Many other ways to reduce pesticide use hazards are available to pesticide users. For instance, the formulation type can make quite a difference as to how much exposure you are getting. Many companies are beginning to formulate products as dry-flowable granules. These aren't meant to be applied to the crop as granules, but to be poured from the bag into the spray tank and dissolved. The fact that they are granules means there is less dust (less product flying around in the air to get on the hands or into the lungs). Recent advances have been made in packaging pesticides to reduce exposure hazards. Glass containers are no longer widely used. Many pesticide packages feature childproof openers.

The method that you choose to store your pesticides greatly influences reduction of use hazards. Guidelines are available on how to properly store pesticides. Remember to have the pesticides in well-vented areas so that toxic fumes don't build up and to store pesticides in a locked area or cabinet. It's also important to have fire extinguishing materials in place near by.

Finally, exposure hazard to pesticides is being reduced by progress in application methods. Controlled droplet applicators (CDAs) are found in many agricultural systems now. These applicators precisely measure and control the droplet size, thereby reducing drift and also reducing the splashing and dripping of large pesticide globs that can occur with older methods. Electrostatic charging of spray particles as they emerge from control droplet and other applicators is a controlled method of reducing hazards. A positive charge is placed on the particle. The particle bends toward and adheres to grounded material such as the plant material being sprayed. The result of this electric charging is less drift and less exposure to the pesticide user.

Thus, we have separated the topic of pesticide safety into three areas and shown how they relate to you as a user. Although acute toxicity and chronic toxicity problems are important and need to be considered, there isn't much you can do about them. The main area where you should increase your skill at using pesticides safely is use-hazard reduction. If you think about hazard reduction when you use pesticides, you can find many easy methods to reduce exposure hazards to any and all pesticide products.

Pesticides and common sense chemistry

Making pesticides work well and avoiding problems associated with their use are two important goals of the material safety data sheet (see Chapter 21) and the pesticide label. Trying to understand how to do this by remembering specific points about this pesticide or that pesticide soon makes one a nervous wreck. It is better to develop useful, common sense ideas about pesticides and their use. Keeping these ideas in mind when using pesticides will be of great value to you.

Pesticides are complex molecules that are constantly associating themselves with other molecules in various chemical reactions. These chemical reactions take place as pesticides mix in a spray tank, as they are delivered through the sprayer onto the plant, as they cover the plant surfaces, as they contact the pest, and as they eventually kill or inhibit that pest. The thing to keep in mind when thinking about pesticide chemistry is that careless use on your part will mean this complicated chemistry will work against you rather than for you.

Pesticide formulations

A pesticide formulation is a pesticide combined with other ingredients in such a way that the product can be used safely to control a pest. There are many kinds of "inert" ingredients included in pesticide formulations. These include spreader-stickers, dispersing agents to keep the molecules apart when mixed with water, wetting agents, solubilizers, stabilizers and substances that simply serve to dilute the active ingredient. There are many different kinds of formulations and many pesticides are formulated in more than one way. The formulation used depends on the market (fruits, vegetables, greenhouse crops) in which that product will be used, the compatibility with application equipment already present in that market, and exactly what crops it's going to be used on. Generally, common formulations include dusts, wettable powders, flowables, soluble powders, emulsifiable concentrates, granulars and fumigants.

Dusts. Dusts are pesticides mixed with finely ground talc, clay or powdered nut shells. As the name implies, dusts are used dry, never with water. Dusts are often used to coat seeds to prevent seed and seedling diseases. They can also be applied directly to foliage. Dust pesticide application tends to be a bit wasteful. It's difficult to make sure the dust reaches the target—only the target—without drifting into other areas. In fact, the rather random dispersal of dusts makes them potentially hazardous, especially in greenhouses.

Wettable powders. These are similar to dusts but contain wetting agents. The particle isn't ground quite as finely as it is in a dust. As the name implies, wettable powders are designed to mix with water to form a semi-stable suspension. Wettable powders are usually more concentrated than dusts. For instance, Captan is formulated as a $7^1/_2$ percent dust, but also can be purchased as a 50 percent wettable powder. Most of the ingredients included in the wettable powder formulation don't come into play until that formulation is mixed with water. A good, wettable powder should mix well by shaking and shouldn't settle out appreciably after standing for 30 minutes. Of course, all wettable powders settle out eventually. Agitation of a wettable powder in a spray tank while it's being applied is generally necessary to prevent dosage problems resulting from settling. Wettable powders can be very abrasive to pumps and nozzles and may clog some low-volume application equipment. Wettable powders leave a visible residue on plant leaf surfaces, particularly when used at 8 ounces or more per 100 gallons of spray. This can detract from the salability of many greenhouse-grown crops.

Flowables. Flowable pesticides are actually very finely ground wettable powders with a wetting agent and water already mixed into the container. They are sold as a suspension that flows much like a water-based paint so the user can mix them with water more easily. Flowable formulations contain more powerful dispersing and wetting agents than most wettable powders. For this reason, flowables won't clog nozzles as much as wettable powders. They will generally mix with water easier than wettable powders, and they are easier to use in low-volume application equipment. Flowables, however, may not be stable in storage. Furthermore, they can leave a more visible residue than wettable powders leave when sprayed onto plant surfaces.

Soluble powders. These are also powdered materials, but these powdered materials dissolve completely and directly in water. Once dissolved, a soluble powder shouldn't require further agitation during delivery. Also, soluble powders aren't nearly as troublesome regarding abrasion and clogging of nozzles. Unfortunately, most pesticides' chemistry prevents them from being formulated as soluble powders. Soluble powder formulations can be hazardous. The pesticide, when dissolved in water, can get into your eyes, nasal passages and be absorbed through your skin more easily.

Emulsifiable concentrates. These liquid pesticide formulations can be mixed with water to form emulsions. The resulting emulsion isn't a solution, but rather a mixture of one liquid dispersed in another. Emulsifiable concentrates mixed with water aren't clear; they are generally milky white because the tiny, emulsifiable, concentrate droplets are dispersed among fine water drops. Many pesticides aren't soluble in water, but are soluble in various oils and organic solvents such as benzene or naphthalene. Thus, they are formulated into emulsifiable concentrates. Emulsifiable concentrates aren't abrasive to spray equipment, but the organic solvents can be somewhat corrosive to washers and other sprayer parts. The main problem with using emulsifiable concentrates is that phytotoxicity (plant damage) from the solvents and emulsifiers can occur. This will be especially hazardous if more than one emulsifiable concentrate is used in any one tank mix.

Granules. These pesticides are prepared by impregnating or binding an active ingredient to a large particle of some sort. The particles may be calcine clay, walnut shells, corn cobs, charcoal or other porous material that has been prepared to a standard size. Many granular pesticides are made to be applied to soil or turf. If applied dry, they don't cling to plants, cutting down on phytotoxicity. They are also safer to apply because there is no wind drift and little dust. There is less danger of skin absorption, but sometimes granules will collect in shoes, pant cuffs and around shirt necks and collars.

Dry flowables. These are a popular, new granular pesticide formulation. They are called flowables because the granules "flow" easily from the container into the tank or measuring device. They are meant to be mixed with water. Upon mixing, the granules disperse into a sprayable suspension. The advantage is they are easier and less hazardous to measure and mix into the sprayer tanks. There is less product waste and less dust generated during preparation.

Preparing pesticides for use

Follow the use instructions on the label and you will be able to take advantage of complicated pesticide chemistry without having to deal with complicated pesticide troubles. For example, always use wettable powders mixed with water. Never use them as dusts. There is generally not a great deal of information concerning the chemistry of the product given on a pesticide label; however, there may occasionally be a pertinent comment. Read the label thoroughly and become as familiar with the product as possible to understand how the product works.

Be careful when mixing pesticides together in one spray tank mix. This most certainly can be done, and it can help save time and increase your effectiveness with pest management practices. When you mix pesticides, give some thought to the process. First, make sure your pesticides are compatible with one another by running a small trial before you apply them. After proper dilution of each, mix them together and note whether their physical structure changes in any way. If you don't get a precipitate or change in color, or have a liquid separation problem, then put this mixture on plant material to see if you have a biological problem. Write down your experiences so that next time you will know what you can and cannot do. Some labels tell you about mixing their products with other pesticides, giving both pros and cons. Avoid using more than one liquid product in a mix.

When mixing one pesticide with another, make sure the concentrated solutions don't come in contact with each other. Dilute the pesticides that you intend to use with a small amount of water in a small bucket before adding to the spray tank. Partly fill the spray tank with water, put the first partially diluted pesticide into the spray tank, mix it well, then add the other partially diluted pesticide. Finally, if an adjuvant or extra spreader-sticker is desired, add this material last after the spray tank is almost filled with the required amount of water. Constantly agitate the liquid in the spray tank during all mixing operations.

Thinking about pesticides as complicated, organic chemicals will help you understand why they may or may not be working. When you deviate from labeled instructions or do things that aren't clearly spelled out as successful, you are playing around with a rather serious chemical situation. It's difficult to predict what might happen when labeled instructions aren't followed. Pesticide use is an important exercise in common-sense chemistry.

Spray adjuvants

A spray adjuvant is one substance added to a second substance that increases the activity of the second substance. Many types of adjuvants are used with pesticides. Some adjuvants increase pesticide adherence to leaf surfaces. These are called stickers. Other adjuvants increase the chemical's systemic uptake. These are called activators. Still other adjuvants reduce drift, serve as anti-foaming agents, or simply wet the leaf so the pesticide is distributed evenly over the surface. The adjuvant label gives some idea about its function.

In spraying to protect plants from insects and pathogenic infections, we generally think of an adjuvant's role as increasing the plant surface coverage. Adjuvants that tend to increase the plant surface coverage are generally called spreader-stickers. Spreader-stickers have two functions. First, they wet the leaf well, creating an even pesticide barrier over the leaf and other plant surfaces. Secondly, they stick the pesticide to plant surfaces and resist the "weathering" activity of the greenhouse environment. This extends and tends to preserve the strength of the residual pesticide.

Another type of adjuvant that is often used for greenhouse crop spraying is one that stabilizes the spray solution's pH. This prevents acid hydrolysis of the product's active ingredient. As discussed below, such "buffering" adjuvants are rarely needed in most greenhouse spray tasks.

An adjuvant could be anionic, ionic, cationic or nonionic. This refers to the charges on the product's molecules. It may be important when using a certain pesticide, but is generally not related to the adjuvant's function. Nonionic adjuvants are widely used on greenhouse crops because they are usually less phytotoxic or damaging to the plants. The key to properly selecting an adjuvant for use is to read the label. Is it sold to be used as you wish to use it? Is it safe to use on the crops you are growing?

Adjuvants can change the chemistry of a pesticide suspension to such a degree that they can contribute to plant damage if used incorrectly. Also, if adjuvants are used at excessive rates, direct plant injury may result.

Growers should follow some general rules when selecting and using adjuvants to minimize plant damage from adjuvants. First, always check a new spray adjuvant on a small group of plants before spraying an entire crop. Check the adjuvant alone and also combined with the pesticides you plan to use on that crop. Do this under conditions that make plants more susceptible to injury. This is likely to be in warm, sunny weather. Normally, plants wouldn't be sprayed under these conditions, but this will be a good test to determine if the spray mixture will be safe to use.

The key to the dilemma of how much adjuvant to use is to remember one rule: *Use the lowest amount needed to wet the foliage sufficiently.* You will have to determine this amount for yourself when using an adjuvant. Start with very low amounts and increase as needed until the foliage begins to wet well using your water and spray equipment. For most greenhouse ornamental plants, we would suggest starting at 1 to 2 ounces per 100 gallons and adding 1 ounce at a time until the desired result is obtained. Probably, 8 ounces per 100 gallons is a maximum amount to use under any circumstance to avoid plant injury.

Sometimes it may become necessary to spray a crop several times in rapid succession. Perhaps a miticide will be applied one day, followed by an insecticide the second day and a fungicide the third day. In such cases, using the same adjuvant in every spray application may produce a build-up and cause plant damage by the time the third spray is applied. In such cases, the adjuvant shouldn't be used in sprays after the initial one. Another solution is to reduce the adjuvant in the subsequent sprays. A third possibility is to make only one application using a tank mix of pesticides. Of course, this can lead to other problems!

Certain adjuvants may increase allergenic problems with pesticides. For example, some growers reported skin reactions among employees using a particular fungicide with adjuvants. There doesn't appear to be a consistent pattern here, so no recommendations can be made, but be aware that this may be a problem.

One final note regarding adjuvants. Some pesticide formulations already contain enough spreader-sticker or other adjuvants. Generally, these will be emulsifiable or liquid products, although at least one wettable powder (Talstar) has a label that states not to use an adjuvant. Use adjuvants only after reading the label on the pesticide(s) you plan to use. The label will often give guidance concerning the use (or non-use) of adjuvants.

Water pH and pesticide performance

It seems that water is being blamed for many things these days. There has been much interest and concern recently about the possible effects of alkaline (above pH 7) water on pesticide effectiveness. Basically, alkaline water can cause the molecules of some pesticides to break down into inactive parts (alkaline hydrolysis). This isn't a simple matter, and the speed of any reaction depends on the pesticide, pH, buffering capacity and temperature. There is no doubt that these things occur. The main question for commercial growers is: How will this affect me? There's no easy answer, but based on the information available to us, it seems that considerable time needs to elapse before significant breakdown of most pesticide products occurs. Therefore, if the pesticide is mixed and used within a few hours and not allowed to stand overnight, there shouldn't be any adverse effects of spray water pH.

Chapter 21

Understanding the
Material Safety Data Sheet

*M*aterial safety data sheets (MSDS) are a fact of life for individuals and businesses dealing with hazardous chemicals. Chemicals are considered hazardous if they fit into several definitions, all of which can be complicated. Briefly, a material is hazardous if it's specifically listed in the law, 29 CFR part 1910, Subpart Z, Toxic and Hazardous Substances (the Z list), assigned an exposure threshold limit value (TLV) by the American Conference of Governmental Industrial Hygienists, Inc. (ACGIH) or determined to be cancer-causing, corrosive, toxic, an irritant, a sensitizer or has damaging effects on specific body organs. For those of us in the greenhouse business, the chemicals involved include pesticides, solvents, oils and lubricants, some paints and possibly some preservatives.

Having to deal with the MSDS may seem unnecessary and trivial. After all, pesticides already have labels on them that alert users to hazards. But if you take the time to learn about the purpose and makeup of an MSDS, you will realize that they are good to have around.

The objective of the MSDS is to concisely inform users about the hazards of the materials you and your employees work with so that you can protect yourselves and respond properly to emergency situations. The law states that all employees who come into contact with any particular hazardous material must have access to the MSDS and be taught to read and understand them.

It's important for you to study the MSDS being sent to you by manufacturers. As you study them, keep in mind several objectives. As you read, imagine how you would respond to emergencies. Think of ways to control day-to-day exposure to materials. Finally, think of how you're going to present this information to your employees. The objective, of course, is for everyone to be aware of hazards, but not to be frightened by them.

Material safety data sheet contents

For pesticides, the MSDS is somewhat redundant of the label, but it isn't identical to the hazard and warning section of the label. The following will be found on every MSDS:

1. The *material's physical properties*.
2. *Fast-acting (acute) toxicities* that make it dangerous to handle.
3. *Protective gear* needed.
4. *First aid* treatment to be provided when necessary.
5. *Procedures* to follow to safely handle spills, fires and day-to-day operations.
6. How to react to *accidents*.

The MSDS is written in English. The MSDS from various suppliers look different, but must, by law, include certain things:

1. Identity of the material and its chemical and common names.
2. Any hazardous ingredients in the formulation.
3. Any cancer-causing ingredients.
4. A list of hazardous physical and chemical characteristics, such as flammability, explosiveness or corrosiveness.
5. A list of acute and chronic symptoms that might occur.
6. Worker exposure limits, the primary entry routes into the body, specific organs likely to be affected and preexisting medical problems that can be made worse by exposure.
7. Precautions and safety equipment needed.
8. Emergency first aid procedures.
9. The date and manufacturer name or the people who prepared the sheet.

MSDS sections

Following are descriptions of MSDS sections and comments to inform you how the information required by OSHA is found and interpreted on a typical good-quality sheet. The MSDS are of varied quality. Beware of any sheets with blank spaces or missing information. Return sheets you judge to be poor quality and request better ones. It's up to the employee to read and follow the instructions on the MSDS. If the sheet isn't properly done, it's you who will suffer.

Heading

The heading gives three things. First, the name, address and phone number of the company that produced the material. Second, it gives the MSDS issue date or most recent revision and third, the name of the material covered.

Section 1. Material identification

This section identifies the material and the supplier. The material name on the MSDS must match the name on the container. If the material has more than one name, each will be listed. The chemical formula may be given. A hazardous-rating fire diamond may appear, giving number ratings for the particular material's degree of flammability, reactivity and health hazard. More specific fire-related information is found in section 4 of the MSDS.

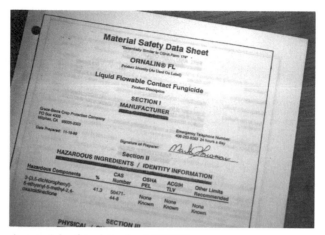

A material safety data sheet (MSDS).

Section 2. Ingredients and hazards

The individual hazardous chemicals in the product and their concentration percentage are listed in section two. If exposure limits have been established for workers over a period of time, they will be shown for each chemical.

LD 50s will be given as well. The phrase "Rat, Oral, LD 50 200 mg./kg." means that 200 milligrams of the chemical per each kilogram of body weight is the lethal dose that killed 50 percent of a group of test rats when the chemical was given to them orally. This group toxicity data helps establish the general degree of hazard to man, but it doesn't tell you how hazardous the chemical is to any one person.

Another important point to keep in mind when reading this section of the MSDS is that exposure to more than one hazardous substance at a time can sometimes be especially harmful. Once in a while, the combined effects of more than one material can prove to be more damaging than the effect of a single material. This is called "synergy." An example of such a potentially damaging combination would be the use of thiram fungicide followed by the consumption of alcoholic beverages.

Section 3. Physical data

Physical data includes items such as a material's boiling point, solubility in water, melting point, evaporation rate, and appearance and odor. Some of this data won't have much relevance, but some of it helps you predict how the material will act and react during handling and use. This enables you to select the correct safety, ventilation and accident-response equipment. This information is particularly hard to understand without some knowledge of chemistry, but if you think carefully about the data given, you can understand how to use it. For instance, if a material has a low boiling point, high vapor pressure, fast evaporation rate and a high percentage of volatility, it's very likely to be an inhalation hazard, and special ventilation or breathing apparatus may be necessary. The higher the temperature, the more active the material.

Section 4. Fire and explosion data

Section 4 of the MSDS indicates equipment to be used by fire fighters and what extinguishing materials work best. You may need to be ready with preplanned response procedures and equipment. In the event of a fire, it's important for firemen to have access to this information.

Section 5. Reactivity data

The information found in section 5 varies greatly from one MSDS to another because of the many different ways that materials may react with one another. It also gives information on storage hazards as they relate to reactivity. For instance, it tells how corrosive the product is on metals. It may tell the temperature and moisture stability. This section states whether the material will polymerize (react with itself), a phenomenon that can cause a rapid buildup of heat and pressure, leading to an explosion. Use this information to guide you in choosing materials for containers, shelving and personal-protection clothing and devices. If the material reacts with natural rubber, you wouldn't want to wear natural rubber gloves.

Section 6. Health hazard information

Section 6 of the MSDS describes all ways the material can gain entry into the body. Acute (immediate) and chronic (long-term) health effects must be stated. If the material is carcinogenic, that fact must be stated. This section also describes medical and first aid treatments for accidental exposure.

Section 7. Spill, leak and disposal procedures

Section 7 advises you how to remedy a spill while safeguarding your health and protecting the environment from further damage. Equally important, the containment and cleanup techniques described will likely reflect compliance with federal, state and local laws and regulations. This can be particularly important for pesticides.

Section 8. Special protection information

Methods for reducing your exposure to a particular hazardous material is described in Section 8. These items are similar to those on pesticide labels; however, they are more general. Many will not be appropriate for the planned use. The methods may include ventilation requirements, breathing apparatus and protective clothing. Instructions for care and disposal of contaminated equipment and clothing are also given.

Section 9. Special precautions and comments

Improper storage and handling of the material can result in the greatest hazard to workers. Proper storage and handling methods are described in section 9, including required labels or markings for the container. Department of Transportation (DOT) policies for handling the material are listed. In addition, section 9 may include comments regarding the material that don't particularly fit the content of other sections of the MSDS.

Chapter 22

Pesticide Use: Common Mistakes, Application Techniques and Record-Keeping

Pesticides are probably our most important tools for combating disease and insect problems on plants. Other health management practices are important, of course, but when the chips are down, most of us rely on a chemical to help solve the problem. People often have trouble managing a problem, even with pesticides. This lack of success often relates to one or more basic mistakes.

The 12 most common pesticide-use mistakes

Following is a list of the most common mistakes. They are presented in the order to follow after deciding to use a pesticide: selection, use and follow-up. They are *not* presented in any order of importance. In fact, you might think of one that is not included at all!

1. Using wrong material because of incorrect diagnosis

We all know that many chemicals are on the market today. Most are specifically used for a certain set of problems. It's difficult to generally categorize pesticides because sometimes a product will be useful for only one or a very limited number of problems. For background information, keep in mind that pesticides often group themselves according to activity against powdery mildews, Botrytis blights and spots, root rots, fungal leafspots, mites, chewing insects, sucking insects and nematodes. There are many examples where this grouping holds true, but there are examples where cross grouping or sub-grouping occurs as well! Diagnose a problem as specifically as you can to avoid choosing the wrong pesticide.

2. Using material incorrectly because of inadequate product knowledge

Even after choosing the correct product because of a correct diagnosis, there can still be problems with it and your decision. The complicated chemistry of pesticides sometimes produces quirks in their behavior. Some, such as the new pyrethroids, may operate best at certain temperatures. Others may break down in certain kinds of light or evaporate into the air too quickly. Most of these "quirks" will be spelled out on labels or in trade journal articles. It always pays to keep your eyes and ears open for this information.

3. Applying the initial application too late

Diseases and insect pests react to crops in sensitive stages and favorable weather conditions. In many cases, reacting to a problem before the rapid problem build-up is the secret to making a chemical work well for you. You can do this by knowing what to look out for, knowing when plants are particularly sensitive to infestation, and knowing what environments favor rapid build-up of particular problems. Rose growers know that young rose foliage is sensitive to mildew infection. They also know that warm days followed by cool nights are favorable periods. Good growers apply preventive mildew sprays when this weather occurs while the roses are coming back from a cutting, cropping or pruning operation. They don't wait until powdery mildew begins to rapidly build up.

4. Improper tank mixes or improper mixing procedure

Tank mixes are preparations of two or more pesticides mixed in the tank for a combined spray. This is widely done in our industry, because it saves time and labor. It often creates problems as well! We are dealing with complicated chemistry here. Just like we tell the kid with a new chemistry set, let's not get carried away. Read labels, try it out on a small scale and mix properly. Don't use more than one liquid or emulsifiable concentrate in a mixture. Mix each pesticide into a separate batch of water and then add to a half-full spray tank slowly, one at a time. Add spreader-sticker last.

5. Using pesticides that have been stored beyond their shelf life

Even though you may be doing everything right so far, watch out for this common error! These complicated, organic chemicals age like wine or whiskey. Unlike wines, however, pesticides lose effectiveness with age. Labels sometimes warn you of this. A material's physical appearance can reflect aging's effect. Is it off-color? Are liquids separated? Are powders lumpy? Does it mix well into water? A few general rules govern this situation. One is that you should never use a product that is over two years old. Another is that you should never allow liquid pesticides to freeze.

6. Not paying attention to weather

Plants go through periods when they are more prone to pesticide burn. Generally, this occurs when they are dry or when the temperature is too warm. Secondly, a hot, sunny exposure right after spraying tends to dry spray droplets too quickly. This causes pesticide concentration in the shrinking droplet and results in a small burn spot when it finally evaporates completely.

7. Applying material at an improper rate per unit area of crop

Labels usually state how much material to mix into the water (often given as "per 100 gallons"). They sometimes don't state how many plants to spray or drench with the 100 gallons! Of course, this varies according to plant type and growth stage. It takes much less spray to treat thinly leafed plants than it does for those with dense foliage canopies. If you're watching your coverage and using a standard high volume technique, you will generally not have rate problems because you're spraying to runoff. This means you wet the foliage just to the point when spray begins to run off the leaf. If you're using a low-volume applicator, you'll use a "rate per area" direction given on the label. It's your responsibility to calculate the area to be treated and the right amount of product with sufficient water to evenly distribute the pesticide solution over the area.

8. Improper coverage of plant material

Pesticides work by getting to the pest or to the plant surface that will soon be infected by the pests. Think about the nature of the pest you are combating. Powdery mildew or spider mites often go uncontrolled because leaf undersurfaces aren't sprayed properly. The type of application equipment you use greatly influences the surface coverage you get.

9. Failing to apply follow-up applications if needed

A single application of something to combat a pest rarely does the trick. Some pests on some leaves escape the spray. Furthermore, the chemical may be ineffective at killing spores or eggs. Thus, they soon germinate or hatch to cause further damage. Finally, since the crop continues to grow, protectant barriers aren't present on new leaves or shoots. Rain and other factors wear away old pesticide barriers.

10. Not keeping thorough records

Doesn't this seem like good old common sense? Suppose something were to go wrong: the plant was damaged or the pest was not alleviated. If you are going to make a good management decision, then you need to change something. This could be a pesticide itself or one component of a mixture, the time of day or the weather, the application method, or the dosage and volume of material applied. You need to know exactly what was done to make a decision. You may need to change more than one thing. If this is the case, change one thing at a time to sort out the problem. Keep these records on each plant. On the next page is a sample pesticide use report you may wish to photocopy and use.

11. Incomplete sprayer cleanup and draining

Again, common sense and deliberate methods enter the picture. Spray equipment is expensive and most pesticides are corrosive. Plant and people safety is at stake. Finally, the "residual" material left in the spray from a previous job may cause a costly mistake on the next crop. Cleanup is part of this job and you should allot time to do it properly.

12. Relying on the chemical alone to do the job

We call this the "spray and pray" school. Health management, whether for you or for your plants, is an integrated, live-right, eat-right, get-sprayed-right philosophy. One without the other won't do. Attention to environmental control, sanitation and plant vigor maintenance forms the basis of health management. If these things are on line, then the pesticides you use will perform well.

Keep good records to avoid mistakes

Every time we view a crop or planting that has been damaged by a pesticide, we are saddened to realize that a bit more attention to detail would probably have prevented the incident. Keeping good records is the best way to prevent mistakes. It's also a pretty good way to improve your chemical control knowledge and skills. Writing down application techniques, dosages and materials used emphasizes their equal importance. Noting the effects of your actions can help you improve methods or evaluate the need for changes.

Pesticide Use Report

Please fill out this report each time a pesticide is applied.

Application date:

Application time:

Reason for application:

Conditions:

Light:

Temperature:

Wind:

Other relevant conditions:

Pesticide(s) used (if more than one, list in order of tank mixing):

Recommended rate:

Amount prepared:

Gallons of water:

Application method:

Area treated (Field No., House No., Bench No., note
    plants not sprayed):

    Area size (square feet or acres):

Total gallons applied:

Total pesticide applied:

Results of application (wait several days after application):

Observation of pest control:

Observation of any damage:

Date results recorded:

Your name:

(Use back of sheet to calculate application amount.)

Pesticide application techniques

Disease and insect pest management is a system that must use many different components. Pesticides (including fungicides, insecticides and acaricides) are important parts of most ornamental plant management programs. This section discusses major aspects of pesticide application and provides information to help you decide what equipment is best for you.

The primary objective of any pesticide application is to deliver the pesticide product to a target in sufficient concentration to control the pest or pathogen involved. Naturally, there are applicator, environmental and plant safety factors to consider, as well as economic factors. The target may be an entire plant, specific plant area, growing medium or the pest (including pathogen, insect or mite). Defining the target is important for proper pesticide application. Is the objective to hit airborne pests? Are the pests on or in leaf surfaces? Are the pests on all plant parts or only in certain areas? Are the pests in the growing medium or under greenhouse benches?

Remember that the pesticide applied and its mode of action are crucial to the success or failure of any application. Does the pesticide have vapor or systemic activity? If so, it will *redistribute* from its point of deposition and reach other areas. Many fungicides are preventive: They prevent a pathogen from becoming established. Most insecticides are eradicative: They eradicate a pest that happens to be present, but don't prevent future infestations. Regardless, pesticide application is a two-step process: deposition and distribution. Deposition is the process of applying the pesticide to the target area, and distribution is the process of getting the material to the correct area in the amounts required to be effective.

Spray drop size is important in determining what kind of deposition and distribution any application method will have. This aspect of pesticide application is often neglected. The pesticide spray may appear to be doing an excellent job of covering the target area, when, in fact, the spray drops are too large (or perhaps too small) to deposit or to be properly distributed on the target.

The following table shows the theoretical coverage obtained with different size spray drops produced from one liter. It's easy to see that small drops will potentially increase coverage. Studies have shown that, with certain insecticides, large numbers of small drops, if distributed properly, will cause higher insect mortality than fewer, larger drops. This allows a lower dosage to be applied and still provide good control. With other insecticides, the smaller drops don't seem to increase efficiency. This information hasn't been established for plant pathogens, but probably is similar to that for insecticides.

Outdoor studies on the relationship between spray drop size and effectiveness against certain targets have shown that very small, aerosol-sized drops, less than 50 microns in diameter, generally are best at hitting small insects and mites. Mid-sized drops, from 50 to 100 microns in diameter, are best at foliage deposition. The largest drops, above 200 microns in diameter, are best in sprays directed at the growing medium surface. Our greenhouse experiments showed that very small drops, such as those produced by the low-volume equipment described below, also deposit sufficient amounts to prevent pathogens from

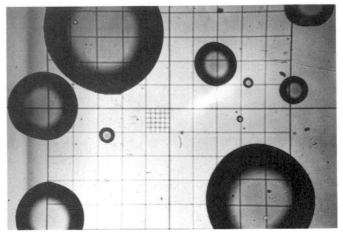

Spray drops as seen through the eye of a needle. The drops may all seem small, but in any spray, some drops are too large and some too small for the best biological effect.

becoming established on foliage or to kill insects and mites on leaf undersides. Residues often were nearly equal to those obtained with high volume sprays. In outdoor situations, drops smaller than 100 microns in diameter may drift excessively and cause pesticide to deposit in areas far away from the intended target. In greenhouses and other protected growing areas, spray drift may be an advantage. In addition to drop size, target characteristics (pest biology, foliage canopy thickness, plant height, bed, floor or bench-grown plants) and equipment characteristics (air movement, flow rate) are important in determining pesticide performance.

Theoretical spray coverage*

Drop diameter (microns)	Drops per square centimeter
10	19,099
20	2,387
50	153
100	19
200	2.4
400	0.3
1,000	0.02

*In drops per square centimeter, applying 1 liter per hectare (2.4 acres) with different spray drop sizes.

High volume sprays

High volume sprays (HV) are the most traditional pesticide application methods in greenhouses and outdoor ornamental crop production, using equipment and methods that haven't changed much over the years. The equipment is available, relatively inexpensive and includes small, hand-held pump sprayers, larger backpack sprayers and large power sprayers. These applications involve mixing a certain quantity of pesticide with a large volume of water and spraying the plants or growing medium to some point of wetness. Water is used in two ways: to dilute the pesticide concentrate, and as a carrier to deliver the material to the target.

When using HV sprays to control small insects, often only about 2 percent to 6 percent of the pesticide applied actually reaches its intended target, with the remaining material lost through evaporation, drift or runoff. HV spray inefficiency in this case may be related to spray drop size. Although some small drops are produced, most of the HV spray volume consists of large drops (greater than 100- to 400-microns diameter), which are very efficient at deposition of product on foliage, significantly increasing spray effectiveness against plant pathogens.

Incorrect dosage sometimes is another reason for reduced HV spray efficiency. It seems obvious that the correct amount of pesticide must be applied for best results; however, determining that amount may be more complicated than it appears. Although changing, many ornamental crop pesticide labels state that the product should be mixed in water (typically 100 gallons) and

High volume spray being applied to roses.

233

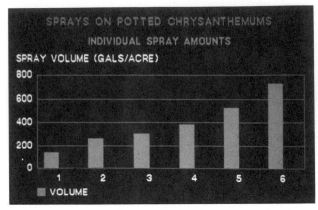

Variations in spray volume applied to the same crop by six different individuals.

"sprayed to run-off," or a similar statement. These directions are generally given without regard to spray volume/crop area, and have different meanings for different applicators. We have seen a five- to six-fold difference in the amount of pesticide applied to the same crop area by different people. A change in labeling on new and re-registered products involves specifying a certain amount of active ingredient and/or spray volume for a crop area. This information will help tremendously in calculating the correct dosage.

High volume, "wet" sprays remain the best, general purpose pesticide application methods. They are time-consuming, often inefficient, wasteful, environmentally undesirable and create re-entry problems; however, the equipment is widely available, relatively inexpensive and remains the only legal way to apply many pesticides. Many pesticide labels for ornamental plant uses are written to effectively prevent using materials at the higher concentrations required by the low volume (LV) sprayers discussed later, even though less total pesticide may be applied in LV sprays.

Also, with proper application techniques (high pressure, getting close to plants and moving the spray nozzle in an arc), foliage canopy penetration and leaf surface coverage are quite thorough and uniform. Therefore, the increased application time is compensated for by good pesticide deposition and distribution. A significant advantage of HV sprays is that they can be used to treat very small areas of a crop (spot treatments). This can be very useful in an integrated pest management program that uses biological control agents, because a spray might be directed to only certain plant or crop parts, avoiding major disruption to the biological control program.

Low volume sprays

Low volume sprays aren't new. They include dusts, smoke generators and total release aerosols, all of which can be useful and effective in many situations. These applications use specialized equipment and require specialized formulations. Other low-volume application equipment may or may not use specialized formulations, and include thermal pulse-jet foggers, mechanical aerosol gener-

A pesticide smoke applicator.
Photo by Molly N. Cline, Monsanto Company.

ators, rotary mist generators and electrostatic sprayers. They eliminate many of the disadvantages of conventional HV applications: They take less time, use less water or oil to dilute and carry the pesticide (no run off, faster re-entry), may use less pesticide and produce most of the spray volume in small spray drops. As mentioned, small drops are potentially more efficient at delivering pesticide to the target. The following summarizes some advantages and limitations of certain types of low volume equipment presently available.

1. Thermal pulse-jet foggers. These applicators have been used for more than 30 years. Originally, foggers were used to apply fumigants in enclosed spaces such as greenhouses. As might be expected, they are very effective at this, given the proper vapor-active pesticide; however, thermal foggers also are effective at applying residual pesticides, often depositing as much pesticide as an HV application. Most models will disperse both liquid or wettable powder formulations. A carrier, or dispersal agent, to be mixed with or used in place of water, may be specified by the manufacturer.

Foggers generate very small drops, usually from two to 50 microns in diameter, that are able to move rather long distances from the applicator. With some of the larger units available, the drops will travel more than 200 feet. Liquid flow rates also vary with unit size. For example, a small model will disperse $2^1/2$ gallons in 30 minutes, and a larger model will disperse 5 gallons in 30 minutes. The area covered with these spray volumes will depend on whether a wettable or liquid formulation is applied. Two-and-a-half gallons will cover about 50,000 square feet with a wettable powder and 76,000 square feet with a liquid formulation. Obviously, foggers aren't used to apply pesticides to small areas in spot treatments. They are designed to treat large areas quickly. When using foggers to apply residual pesticides, especially wettable powder formulations, it's very important to apply by aiming the spray over the crop at about a 30 degree angle. If this isn't done, heavy deposition in areas immediately in front of the fogger will result, causing excessive dosage and plant injury. Some growers who have expanded metal greenhouse benches have made applications from below the plants, allowing the fog to rise up through the plant canopy.

Thermal pulse-jet fogger.

Those who have used this technique report good results, but have provided no data to support these observations.

When applied from above the crop, the pesticide distribution within the plant canopy and leaf undersides may be poor. Thus, using thermal pulse-jet applicators with certain pesticide/pest combinations won't be successful; however, using a pesticide that redistributes after application through systemic or vapor action can be very effective.

2. Rotary mist applicators. Originally called controlled droplet applicators (CDA) or rotary atomizers, these machines have been quite useful in many situations. They aren't widely available for use in greenhouses in the United States, but are used outdoors for such diverse purposes as mosquito and weed control. These are portable sprayers that disperse pesticides by directing the flow onto rapidly rotating, notched discs. This causes the liquid to break up into small drops, 10 to 60 microns in diameter. A fan behind the disc propels the spray toward the target, creating a turbulent air stream. Depending upon the crop to be treated, one to four quarts are dispersed per acre. As with the amount of liquid dispersed, the time to treat a given area will vary. Workers carry sprayers through the crop with a walking speed of one pace per second. The spray is aimed ahead about 10 feet, directed at the crop, and moved up and down to ensure proper coverage. Obviously, bedding plants can be treated more rapidly than roses. These devices aren't space sprayers, and must be aimed at the crop for best results. Pesticide deposition and distribution is quite good, but foliage canopy penetration and leaf underside coverage varies significantly, depending upon crop type: rose, bench-grown potted plants or bedding plants on the floor.

Most pesticides applied in rotary mist applicators are specially formulated for this use. Only one pesticide, Rotospray Resmethrin, is currently registered for use with this equipment in the United States, but some growers have had success with their own mixtures. To do this, calibrate each mixture to ensure the correct dosage.

A rotary atomizer.

3. Mechanical aerosol generators. These sprayers are designed to be operated unattended and use air pressure supplied by an air compressor to break up the spray liquid into small, fog-sized drops. Air is also the main method of moving the spray around the greenhouse and onto the foliage. A fan on each unit helps propel the spray drops, but much of the spray movement is accomplished by the greenhouse air movement system: horizontal air flow or overhead convection tubes. Applications can be made into long polyethylene tubes with small holes along the tube length to help distribute the pesticide. Depending on the model, a single sprayer will treat an area of more than 30,000 square feet, using less than 4 gallons of liquid. You can place the spray mixture in the tank, and set a timer to make the application when no workers are present. The liquid flow rate is quite slow, about 2 fluid ounces per minute, so the treatment may take several hours. After the application is complete, ventilate the greenhouse before any workers re-enter.

Our results, using fluorescent tracer material as well as nonsystemic insecticides, have shown that pesticide deposition and distribution are quite good. Although tracer studies indicate that deposition on upper leaf surfaces was much higher than on undersides, control of first-instar, greenhouse whitefly nymphs on poinsettia with bifenthrin was excellent. Similar results have been obtained against two-spotted spider mites and melon aphids. Performance likely will vary, depending on the crop, pest and pesticide applied. Very few studies have been conducted using fungicides. One possible safety problem with these sprayers is off-target deposition. Because the pesticide circulates

Coldfogger pesticide applicator.

throughout the greenhouse, it's deposited on all surfaces, such as fans, benches, walls and the sprayer itself.

4. Electrostatic applicators. Electrically charging spray drops to achieve better coverage of lower leaf surfaces and reducing spray drift is an idea that has been around for many years. We have evaluated two types of electrostatic low volume applicators, one type with air-assistance and one without. One unit without air-assistance required special formulations to properly charge spray drops. The spray was dispersed by the electrical charge and by gravity. Although foliage deposition was quite good, distribution was restricted to upper leaves. Another unit without air-assistance was a prototype rotary mist applicator that used a rotating disc to disperse charged drops. No special formulations were necessary; however, lack of air-assistance resulted in poor foliage canopy penetration.

Aerosol generator.

Air-assisted electrostatic sprayer.

Recently, three air-assisted electrostatic sprayers became available in the United States. Two sprayers are hand-held and the third is operated unattended. All use similar spray nozzle technology and require an air compressor to atomize and transport the spray. These sprayers don't require special formulations. We evaluated both hand-held models, one with a single nozzle and the other with twin nozzles, and found them to be significantly more effective than the electrostatic sprayers without air-assistance. The air-assistance is very important in penetration of the plant foliage canopy. Both sprayers produce drops with a volume median diameter of about 30 microns. In our experiments, the twin nozzle sprayer had a liquid flow rate of 10 fluid ounces per minute, and applied from 4 to 12 gallons per acre, depending on crop size and walking speed. The single nozzle sprayer had a liquid flow rate of about 2^1/2 fluid ounces per minute, producing a total spray volume of 4 gallons per acre.

Both devices often had a "plant position" effect; that is, pest control sometimes was better on plants nearest the sprayer nozzle. This effect varied with the pesticide and sprayer, but was more obvious with the single-nozzle sprayer, which had less air movement and liquid flow rate. Subsequent models have been modified to reduce this problem. With both sprayers, better control was obtained with applications from both sides of the greenhouse bench, but the twin-nozzle model has more air flow and should provide good coverage and deposition from only one side of the plant bench or bed. As with any new type of equipment, learning how to make the best use of electrostatic sprayers will take time. Using both sprayers to apply bifenthrin at application rates well below those used in HV sprays resulted in excellent whitefly nymph control on poinsettias arranged in several positions across a greenhouse bench. Charged sprays were generally more effective than uncharged sprays, but not always. The time required to make an application on most bench or ground bed crops should be about half that required by a HV spray over the same area. Because LV sprays don't wet the foliage, re-entry can occur after adequately ventilating the greenhouse. Off-target deposition is also less with electrostatic sprayers, compared with aerosol generators.

The future of low volume application

Low volume applications have the potential to significantly reduce the amount of pesticide applied while maintaining good pest control. The basic facts, however, are that technology is well ahead of legality in many cases. Any application method not prohibited on the label can be employed to apply that pesticide. If a specific dilution (for example, 1 pound per 100 gallons) is required, however, this effectively prevents low-volume applications, which must be made at very high concentrations (10 to 25 times greater than HV sprays) to be effective. Most labels are written this way, but changes are occurring. Some pesticide companies are modifying their labels to allow for low-volume application. More insecticides are labeled for use in low volume sprays than fungicides.

Each greenhouse operation should have both a low-volume sprayer and a high volume sprayer. The specific kind of low volume sprayer will depend on many factors, including greenhouse size, whether the operation includes many separate houses or one large, interconnected range, and the crops produced. For example, you may have a greenhouse where smoke generators or total release aerosols will be most practical. Others may have very a large, open greenhouse that requires thermal foggers or aerosol generators for best results. No single sprayer is best for everything. The three most important factors in pesticide application, in addition to equipment are: pesticide, pest or disease biology, and application interval.

Chapter 23

Putting It All Together:

Writing Plant Protection Programs

*T*he primary goal of reading a book like this is to be able to keep the crops you are producing healthy and pest free. A secondary goal is to be able to react quickly and properly when a problem arises. You accomplish the first goal by becoming aware of potential problems and planning operations to prevent their occurrence or development. You accomplish the second goal by correctly diagnosing the problem, determining its primary and secondary causes and deciding on and carrying out reactive health management (curative) programs.

Many people think that the reactionary actions outlined above are the main objectives of learning about diseases and pests. A problem occurs, the books are read, curative actions are taken. This sort of "after the fact" knowledge can often save growers from disaster, but it more often results in a sad case of having to learn from hindsight what should have been done in the first place. This chapter shows you how to write plant protection programs to keep plants healthy. It's really a three-step planning process. You must initially think about the crop you are growing. If it's one crop, such as poinsettia or geranium, it's not too difficult to determine what diseases and insects might be present. The information in this book can help you a bit, but more detailed plant disease and pest dictionaries are available. Buy one and become familiar with the potential disease and insect problems of the plant types you will be growing.

The second step in the plant protection planning process is the most difficult. After you have this great list of potential disease and insect problems in front of you, decide which are sufficient threats to warrant a preventive plant protection procedure. Actually, you must put each disease or pest into one of the following three categories: 1. Definitely a problem unless protective measures are taken. 2. So rare that it can be discounted as long as things go normally. 3. A problem that is relatively uncommon but so disastrous when it occurs that you need to be always watchful for it and ready to take quick action if it shows up.

Where do you get help in getting through this second step? The crop programs that follow represent our opinion of problems in the first category. Notice as you change from crop to crop, many problems appear again and again. These are the basic plant protection elements of our industry.

You may not agree with our inclusion of a particular problem on a particular plant. At least, you may not agree that the problem is one that warrants the protective activity that we have suggested. There is nothing wrong with your feeling this way. Your knowledge and experience may have taught you that the problem isn't one of great importance in your growing operation. You may, however, have simply been lucky. Also, it may be that some other health management practice that you routinely do just happens to prevent the problem we have mentioned. Watch out for these pitfalls!

This brings us to the third step in writing a plant protection program: actually writing down a plan. Once you decide that a particular disease or pest warrants some sort of protective practice on a crop you are growing you must decide which practice to use. In the various chapters of this book, we have attempted to show you how integrated management practices are the cheapest and most successful ways to prevent insect or disease problems.

The best way to write a plant protection program is to follow the production steps of the crop in question. This is true for several reasons. Many pathogens and pests prefer to attack a crop in one of its maturity stages. Treatments and practices used will depend somewhat on the crop growth stage. For instance, pesticide damage is more common on young, tender seedlings and on open flowers. Thus, spraying decisions will depend on crop age.

We have suggested some plant protection practices in the following paragraphs. How many of them can you use? How cheaply can you do them while still doing them effectively? There is only one hard and fast rule. Don't write plant protection programs based solely on pesticide use. They simply won't work. Appreciate the role of secondary factors such as environmental manipulation or sanitation and the need to constantly observe your crop and scouting for pests or diseases. Include these practices in your plant protection programs.

Plant protection program samples

Poinsettia plant protection program

Crop stage	Plant protection practice	Purpose
Propagation	Place potted stock plants on sterilized surface of raised bench for growing on.	**Diseases prevented:** Bacterial canker Bacterial soft rot Pythium root rot Rhizoctonia basal stem rot Thielaviopsis black root rot
	Space plants well and provide for good air circulation.	**Diseases prevented:** Bacterial soft rot Botrytis blight
	Organize a record-keeping system for pesticide applications, pest counts on sticky traps and pest counts on plants.	**Prepare for insect control activities**
	If stock plants are to be grown from cuttings, root the cuttings on the sterilized surface of raised benches. Sticking cuttings into steam-treated or fumigated propagating medium in sterilized flats or containers, or use sterile propagating blocks following strict sanitation procedures. Fill bench or flats with propagating medium. Place propagating tools and flats on bench, cover and steam at 180 F for one-half hour at coolest point **OR** fumigate all media and materials with methyl bromide.	**Diseases prevented:** Pythium root rot Rhizoctonia basal stem rot Thielaviopsis black root rot
	Hang hose so nozzle doesn't touch floor.	**Disease prevented:** Pythium root rot
	Don't allow dust to contaminate propagating medium or sterile propagating blocks.	**Diseases prevented:** Rhizoctonia basal stem rot Thielaviopsis black root rot

Wash hands thoroughly before taking cuttings from stock plants. Remove cuttings from stock plant with a knife sterilized by dipping in 70 percent alcohol. Change knives between stock plants **OR** use a series of knives, changing between each stock plant. Soak the knives for at least five minutes in a quaternary ammonium salt or phenolic sanitizer before reuse.

Place cuttings immediately into a new plastic bag or sterile container. Apply rooting hormones with a powder duster. Don't dip cuttings into any solution.

Stick cuttings directly into steam-treated propagating medium or sterile blocks from the plastic bags or sterile containers.

If fungus gnat activity appears, apply a drench application of an appropriate insecticide.

If root or basal stem rot appears while the cuttings are rooting, drench affected plants plus a 2- to 3-foot area around the affected cuttings with fungicides effective and labeled on poinsettia for Rhizoctonia, Pythium and Thielaviopsis. This may require mixing two fungicides. Apply one pint of the suspension per square foot of bench area. Make one application only.

If Botrytis blight appears, spray affected area with a registered fungicide or fumigate with Exotherm Termil. Turn off mist early enough in the evening to allow the leaves to dry prior to nightfall.

Diseases prevented:
Bacterial canker
Bacterial soft rot

Control fungus gnats

Control root or basal stem rot complex

Control Botrytis blight

Prepare for planting

If bacterial soft rot appears, remove affected cuttings at once. If possible, lower the temperature. Lower the nitrogen rate being given to the stock plants to correct the problem for the next batch of cuttings.

Control bacterial soft rot

If using biological control, make sure Encarsia parasite shipments have been scheduled.

Plan for biological whitefly control

Inspect cuttings for presence of whitefly eggs or nymphs. If significant numbers are found, plan to make three to four applications of an appropriate insecticide as soon as plants are established. Use insecticides less harmful to parasites if they are to be introduced.

Keep whitefly populations low

When cuttings have rooted, remove with clean hands and place in sterile containers.
Hang hose so nozzle doesn't touch the floor.

Diseases prevented:
Bacterial canker
Pythium root rot
Rhizoctonia basal stem rot
Thielaviopsis root rot

Hang yellow sticky traps near plant tops, and place some traps, sticky side up, on growing medium surface. Use up to four traps per 1,000 square feet for best results. If this trap number is too high for record-keeping, use a number appropriate for you. When detecting whiteflies, plant inspection is often more effective than trapping.

Detection of whiteflies and fungus gnats

Prepare potting mix for planting. Allow for good aeration. Adjust media pH to 4.5 to 5.5.
Fill pots with growing mix, place on greenhouse bench. Cover bench, pots and potting tools and steam at 180 F for one-half hour at coolest point **OR** fumigate or steam potting mix in a clean area a few days before use. Store it in an area where dust won't contaminate it. If a pathogen-suppressive potting mix is to be used, make sure it is fresh and in unopened and undamaged bags.

Diseases prevented:
Pythium root rot
Rhizoctonia basal stem rot
Thielaviopsis root rot

Planting

Plant rooted cuttings directly from sterile container, flats or rooting blocks. Always wash hands thoroughly before planting rooted cuttings or pinching growing plants.

Diseases prevented:
Bacterial canker
Pythium root rot
Rhizoctonia basal stem rot
Thielaviopsis root rot

Growing the plants

After the first watering, drench with soil fungicides, using a combination of products to protect against both Rhizoctonia and Pythium. Apply one-half pint of suspension per 6-inch pot, one application only.

Diseases prevented:
Pythium root rot
Rhizoctonia basal stem rot

If using biological control, begin Encarsia introductions during the first two weeks after potting. Be very careful with pesticide use.

Biological control of whiteflies

Immediately after potting, begin a weekly plant scouting program for whiteflies, mealybugs and aphids. Inspect plants in all greenhouse areas, but pay special attention to areas near open vents and doors. Inspect sticky traps weekly. Keep records. If whiteflies are detected and biological control isn't being used, begin a pesticide application program.

Insect detection and population monitoring

If yellow traps indicate significant fungus gnat activity during the first week or two, apply an insecticide drench.

Fungus gnat management

Plants panned in August or September may require a second fungicide drench about October 15.
Water thoroughly, but only when needed.

Diseases prevented:
Pythium root rot
Rhizoctonia basal stem rot

Make sure all high-volume sprays are applied before bracts begin to show color. If pesticides are necessary after bracts show color, use smokes, fogs or aerosols.

General insect management

As crop matures, guard against excessive greenhouse humidity. Never water late in the day. Keep air moving over the foliage at all times. If damp, rainy weather is a problem, fumigate the greenhouse with Exotherm Termil, but do so only before the bracts have begun to show color.

Disease prevented:
Botrytis blight

As the crop matures, make sure the nitrogen levels are decreased, especially if the weather is cloudy and warm. Ventilate vigorously to prevent the temperature from getting into the 80s.

Disease prevented:
Erwinia stem canker

Geranium plant protection program

Crop stage	Plant protection practice	Purpose
Propagation	Purchase culture-indexed and virus-indexed cuttings for stock plants. Pot the cuttings into new containers in sanitized media (see below).	**Diseases prevented:** Bacterial stem rot and leaf spot Verticillium wilt Various viruses
	Organize a record-keeping system for pesticide applications, pest counts on sticky traps and pest counts on plants.	**Record keeping for effective insect control**

Use yellow sticky traps, placed horizontally on the growing medium surface, to monitor fungus gnat activity. If traps are catching more than a few adults per week during the first month, apply an insecticide drench.

Fungus gnat management

Flower stock plants only to verify variety. Remove and destroy other flowers in the bud stage.
Vent and heat at sundown to reduce the relative humidity during spring and fall. Keep air moving over the plant at all times. Never water late in the day.

Diseases prevented:
Alternaria leaf spot
Botrytis blight

One month before taking first cuttings, thermal dust stock plants each week with Exotherm Termil or spray with Daconil 2787, Ornalin or Chipco 26019 according to labeled directions.

Disease prevented:
Botrytis blight

Hang hose so nozzle doesn't touch greenhouse floor.

Diseases prevented:
Bacterial fasciation
Fusarium blackleg
Pythium blackleg

Don't allow visitors or employees not connected with the crop to go into the area where the stock plants are growing.

Diseases prevented:
Bacterial stem rot and leaf spot
Various viruses

Propagation Select a raised propagating bench away from other areas of geranium production.

General disease control

Place propagating medium and tools on propagating bench and steam for one-half hour at 180 F at coolest point **OR** use sterile propagating blocks **OR** use fumigated propagating medium with tools sanitized in a liquid sanitizing agent.

Diseases prevented:
Bacterial fasciation
Fusarium blackleg
Pythium blackleg
Rhizoctonia root and stem rot

Wash hands thoroughly and rinse in a diluted sanitizing agent. Break cuttings from stock plants, place in clean, sterile flats lined with new newspapers **OR** remove cuttings with a knife disinfested by dipping in 70 percent alcohol or use a series of knives, changing between each stock plant. Use enough knives so that they can soak for at least 10 minutes in a quaternary ammonium salt or phenolic sanitizer before reuse. Place cuttings in clean, sterile flats lined with new newspapers.

Diseases prevented:
Bacterial stem rot and leaf spot

Take cuttings from flats and stick directly in steamed propagating medium in benches, flats, trays or pots. Don't dip cuttings into any solutions. If rooting hormone is to be used, dust it onto the cuttings. Don't allow dust to contaminate propagating medium or sterile propagating blocks.

Diseases prevented:
Bacterial fasciation
Bacterial stem rot and leaf spot
Fusarium blackleg
Pythium blackleg
Rhizoctonia root and crown rot

Turn off mist early enough in the evening to allow leaves to dry prior to nightfall.

Disease prevented:
Botrytis blight

Planting

Prepare planting mix, adding sphagnum peat moss, bark, Perlite or other amendments before steaming.
Fill pots with soil mix, place on the production bench. Steam pots, bench and potting tools at 180 F for one-half hour **OR** fumigate or steam the batch of potting mix in a clean area a few days before use. Store it in an area where dust will not contaminate it. Use only new or sanitized pots **OR** use a sanitary media directly from a plastic bag.

Diseases prevented:
Bacterial fasciation
Bacterial stem rot and leaf spot
Fusarium blackleg
Pythium blackleg
Rhizoctonia root and crown rot

Wash hands thoroughly; rinse in a diluted sanitizing agent.

Diseases prevented:
Bacterial stem rot and leaf spot

If pathogen-free rooted cuttings are purchased, plant from shipping container directly into steam-treated planting mix in pots on production bench **OR** remove rooted cuttings from propagating bench, and place in clean sterile flats lined with new newspapers. Plant from flats directly into steam-treated planting mix in pots on production bench.

Diseases prevented:
Bacterial fasciation
Bacterial stem rot and leaf spot
Fusarium blackleg
Pythium blackleg
Rhizoctonia root and crown rot

After the first watering, drench with soil fungicides, using a combination of products to protect against both Rhizoctonia and Pythium. Apply one-half pint of suspension per 6-inch pot, one application only.

Hang yellow sticky traps near plant tops, and place some traps, sticky side up, on growing medium surface. Use up to four traps per 1,000 square feet for best results. If this trap number is too high for record keeping, use a number appropriate for you. For whitefly detection, weekly plant inspection is often more effective. Thrips detection is usually best using traps.

Detection of whiteflies, thrips and fungus gnats

If yellow traps indicate significant fungus gnat activity during rooting, an insecticide drench is required.

Control fungus gnats

Growing plants Hang hose so that nozzle doesn't touch the greenhouse floor. Avoid overhead watering.

Disease prevented:
Pythium blackleg

Wash hands thoroughly and rinse in diluted sanitizing agent before pinching plants.

Disease prevented:
Bacterial stem rot and leaf spot

Remove and destroy flowers before they shatter and petals fall on leaves. If weather is damp or rainy, thermal dust or spray weekly with a fungicide for Botrytis.

Vent and heat at sundown to reduce relative humidity. Space plants for good ventilation. Water only in morning, avoid overwatering when cool or cloudy. Keep air moving over the plants at all times.

Disease prevented: Botrytis blight

Keep whiteflies and thrips under control. An established infestation of either pest group will require four to six insecticide applications at five- to seven-day intervals to obtain control.

Control whiteflies and thrips

Watch for chewing or leaf-rolling injury caused by caterpillars. Cutworms will be in the growing media at the plant base during daylight hours. Apply a pesticide late in the day because these insects are active at night.

Caterpillar management

Easter lily plant protection program

Crop stage	Plant protection practice	Purpose
Prepare for planting	Prepare planting mix. Allow for good aeration. Put mix, pots and potting tools on raised bench; cover with steam cover. Steam this and other benches to be used for lily production at 180 F for one-half hour at coolest point **OR** fumigate or steam potting mix in a clean area a few days before use. Store so dust won't contaminate it. If a pathogen-suppressive mix is to be used, make sure it is fresh and in unopened, undamaged bags. Hang watering hose so nozzle doesn't touch greenhouse floor.	**Diseases prevented:** Pythium root rot Rhizoctonia root rot

	Organize a record-keeping system for pesticide applications, pest counts on sticky traps and pest counts on plants.	**General insect management**
Planting	Plant bulbs directly from shipping container; pot directly onto a steam-treated or otherwise sanitized production bench. Don't set pots on greenhouse floor.	**Diseases prevented:** Pythium root rot Rhizoctonia root rot
	After the first watering, drench with soil fungicides using a combination of products to protect against both Rhizoctonia and Pythium. Apply one-half pint of suspension per 6-inch pot.	
	Using a hand lens, examine some bulbs for bulb mites. Low numbers won't be seen easily. If bulb mites are a concern, soak bulbs in a suspension of Kelthane 35 percent wettable powder and water (1 $^1/_3$ pounds per 100 gallons) for 30 minutes.	**Bulb mite management**
Growing plants	Water plants thoroughly, but only when needed.	**Diseases prevented:** Pythium root rot Rhizoctonia root rot
	Keep nitrogen levels high, relative to potassium and phosphorus, until plants are ready for market.	
	Drench with soil fungicides monthly with a combination of products to protect against Rhizoctonia and Pythium. Apply one-half pint of suspension per 6-inch pot.	
	Inspect plants regularly to detect aphids. Pay special attention when buds appear. If aphids are found, make at least two pesticide applications seven days apart.	**Aphid management**

Bedding plant protection program

Crop stage	Plant protection practice	Purpose
Prepare for sowing seed and planting seedlings	Use only trays or flats that are new or that have been sanitized with a quaternary ammonium chloride or phenolic compound. Prepare planting mix, adding sphagnum peat moss, bark, Perlite or other amendments before steaming. Fumigate or steam the potting mix in a clean area a few days before use. Store in an area where dust will not contaminate **OR** use a sanitary media directly from a plastic bag. Adjust media pH to 4.5 to 5.5.	**Diseases prevented:** Damping off Pythium root rot Rhizoctonia basal stem rot Thielaviopsis root rot
	Organize a record-keeping system for pesticide applications, pest counts on sticky traps and pest counts on plants.	**Record keeping for general insect control**
Producing plugs or seedlings	Sanitize seeder machines each week or between each batch of seed.	**Diseases prevented:** Damping off Pythium root rot Rhizoctonia basal stem rot Thielaviopsis root rot
	If sowing in rows, sow thinly.	**Disease prevented:** Botrytis blight and damping off
	Treat trays of petunias, pansies or vincas with a benzimidazole fungicide immediately after seeding.	**Disease prevented:** Thielaviopsis black root rot
	Germinate seed under high humidity and at the correct temperature for the plant species.	**General disease control**
	When seedlings emerge, remove trays from high humidity germination chamber and treat for water mold diseases. Also treat with benzimidazole fungicide if not done earlier.	**Diseases prevented:** Damping off Pythium root rot Rhizoctonia basal stem rot Thielaviopsis root rot

If weather is damp and cloudy, spray seedlings lightly with Daconil 2787 or Chipco 26019.

Disease prevented: Botrytis blight and damping off

Hang yellow sticky traps near plant tops, and place some traps, sticky side up, on growing medium surface. Use up to four traps per 1,000 square feet for best results. If this trap number is too high for record-keeping, use an appropriate number. For whitefly detection, weekly plant inspection is often more effective than trapping. For thrips detection, traps are usually better than plant inspection.

Detection of whiteflies, aphids, thrips and fungus gnats

Transplanting and growing plants

Fumigate or steam the potting mix in a clean area a few days before use. Store so dust won't contaminate it. Use only trays or flats that are new or that have been sanitized with a quaternary ammonium chloride or phenolic compound.
Adjust media pH to 4.5 to 5.5.
Place flats on sanitized surfaces only for growing on.
Keep walkways clean and free of dust. Sanitize monthly.
Do not overfertilize, especially with ammonia nitrogen products.

Diseases prevented: Damping off Pythium root rot Rhizoctonia basal stem rot Thielaviopsis root rot

Immediately after potting, begin a weekly plant scouting program for whiteflies, aphids and spider mites. Inspect plants in all greenhouse areas, but pay special attention to areas near open vents and doors. Inspect sticky traps weekly for whiteflies, fungus gnats and thrips. Keep records. If an infestation of one or more pests is detected, begin a pesticide application program. The proper material and number of applications will depend on the pest and host plant.

General insect detection and population monitoring

After the first watering of transplanted seedlings, drench with soil fungicides, using a combination of products to protect against both Rhizoctonia and Pythium. Apply 100 gallons of suspension per 800 square feet of flats or over whatever area of flats it takes to water the suspension well into the planting media. Make one application only.

Diseases prevented:
Damping off
Pythium root rot
Rhizoctonia basal stem rot
Thielaviopsis root rot

Vent and heat at sundown to reduce relative humidity. Space plants for good ventilation. Water only in morning; avoid overwatering when cool or cloudy. Keep air moving over the plants at all times.
In cool, cloudy weather, spray with Daconil 2787, Ornalin or Chipco 26019 according to label directions. Avoid fungicide damage to plants in flower.

Disease prevented:
Botrytis blight and damping off

Appendix I

Where to Write for This or That

Arizona

Chemonics Labs
734A E. South Pacific Dr.
Phoenix, AZ 85034
(602) 262-5401

California

A & L Laboratories
1311 Woodland Ave.
Modesto, CA 95351
(209) 529-4080

DANR Analytical Lab
Univ. of California
Davis, CA 95616
(916) 752-0147

Morse Labs
1525 Fulton Ave.
Sacramento, CA 95825
(916) 481-3141

Soil & Plant Lab
P.O. Box 153
Santa Clara, CA 95052
(408) 727-0330

Soil & Plant Lab
1594 N. Main St.
Orange, CA 92667
(714) 282-8777

Florida

A & L Laboratories
1301 W. Copans Rd.
Building D, Ste. 8
Pompano Beach, FL 33064
(305) 972-3255

Scientific Laboratories Inc.
7544 Lawrence Rd.
Lantana, FL 33464
(407) 547-5776

Soil Testing Lab
Wallace Bldg.
IFAS 0542
Univ. of Florida
Gainesville, FL 32611
(904) 392-1950

Georgia

Plant and Water Analysis Lab
2400 College Station Rd.
Univ. of Georgia
Athens, GA 30605
(404) 542-5350

Illinois

Top Soil Testing Service Co.
27 Ash St.
Frankfort, IL 60423
(815) 469-2530

Indiana

A & L Laboratories
3505 Conestoga Dr.
Ft. Wayne, IN 46808
(219) 483-4759

Michigan

Fishbeck, Thompson, Carr
& Huber
6090 E. Fulton
P.O. Box 211
Ada, MI 49301
(616) 676-2666

Soil Testing Lab
A81 Plant & Soil
  Science Bldg.
Michigan State Univ.
East Lansing, MI 48824
(517) 355-0218

Nebraska

A & L Laboratories Inc.
13611 B St.
Omaha, NE 68144
(402) 334-7770

Harris Laboratories Inc.
624 Peach St.
Lincoln, NE 68502
(402) 476-4091

New Jersey

General Testing Corp.
85 Trinity Place
Hackensack, NJ 07601
(201) 488-5242

Industrial Corrosion Mgt.
P&P Laboratory Inc.
2025 Woodlynne Ave.
Woodlynne, NJ 08107
(609) 962-6611

New York

Cornell Nutrient
  Analysis Lab
20 Plant Science
Cornell Univ.
Ithaca, NY 14853
(607) 255-4532

North Carolina

Compuchem
P.O. Box 12652
Research Triangle
 Park, NC 27709
(919) 549-8263

Soil Testing Lab
Agronomy Div.
2109 Blue Ridge Rd.
Raleigh, NC 27607
(919) 733-2656

Ohio

Brookside Research Lab
308 S. Main St.
New Knoxville, OH 45871
(419) 753-2448

OARDC
Research & Ext.
 Analytical Laboratory
Wooster, OH 44691
(216) 263-3700

Pennsylvania

Grace-Sierra
6656 Grant Way
Allentown, PA 18106
(215) 395-7104

Tennessee

A & L Laboratories
411 N. Third St.
Memphis, TN 38105
(901) 527-2780

Texas

A & L Laboratories
302 34th St.
P.O. Box 1590
Lubbock, TX 79408
(806) 763-4278

Ext. Soil Testing Lab
Soil and Crop Science Bldg.
Texas A & M
College Station, TX 77843
(409) 845-4816

Virginia

A & L Laboratories
7621 White Pine Rd.
Richmond, VA 23237
(804) 743-9401

Washington

Soil & Plant
P.O. Box 1648
Bellevue, WA 98009
(206) 746-6665

EXTENSION SPECIALISTS IN ENTOMOLOGY FOR ORNAMENTAL PLANTS IN THE U.S. AND CANADA

(Many of the specialists listed here have additional responsibilities. Some listed for Canada have general greenhouse responsibilities and are not necessarily involved with insect or mite problem-solving. These individuals, however, can facilitate obtaining the correct information.)

Alabama

Patricia P. Cobb
Extension Hall
Auburn Univ.
Auburn, AL 36849-5413
(205) 826-4940

Alaska

Wayne Vandre
Univ. of Alaska
2221 E. Northern Lights Blvd.
Ste. 240
Anchorage, AK 99508
(907) 279-6575

Arizona

Dave T. Langston
4341 E. Broadway
Univ. of Arizona
Phoenix, AZ 85040
(602) 255-4456

Robert L. Smith
Entomology Dept.
Univ. of Arizona
Tucson, AZ 85721
(602) 621-7167

Arkansas

Donald R. Johnson
Cooperative Ext. Service
Box 391
Univ. of Arkansas
Little Rock, AR 72203

California

Richard S. Cowles
Dept. of Entomology
Univ. of California
Riverside, CA 92521
(714) 787-4737

Colorado

Whitney S. Cranshaw
Dept. of Entomology
Colorado State Univ.
Fort Collins, CO 80523
(303) 491-6781

Connecticut

M.H. Brand, R.J. Macavoy,
 E. L. Marotte
Plant Science Dept.
Univ. of Connecticut
1376 Storrs Rd.
Storrs, CT 06268
(203) 486-3435

Delaware

Dewey M. Caron
Dept. of Entomology and
 Applied Ecology
Univ. of Delaware
Newark, DE 19717
(302) 451-2526

District of Columbia

M.S. Khan
DC Ext. Service
901 Newton Street, NE
Washington, DC 20017
(202) 576-7419

Florida

Russell F. Mizell
AREC
Univ. of Florida
Monticello, FL 32344
(904) 997-2596

Donald E. Short
Dept. of Entomology
 & Nematology
Univ. of Florida
Gainesville, FL 32611
(904) 392-1938

Georgia

Randall D. Hudson,
 William G. Hudson
Box 1209
Univ. of Georgia
Tifton, GA 31793
(912) 386-3424

Ronald D. Oetting
Georgia Experiment Station
1109 Experiment Street
Griffin, GA 30223-1791
(404) 228-7288

Beverly Sparks
Barrow Hall
Univ. of Georgia
Athens, GA 30602
(404) 542-1765

Guam

Cooperative Ext. Service
College of Ag. and
  Life Sciences
Univ. of Guam, UOG Station
Mangilao, GUAM 96923
FAX 671-734-2575

Hawaii

Ronald F. L. Mau
3050 Maile Way
Room 310
Univ. of Hawaii at Manoa
Honolulu, HI 96822
(808) 948-7063

Arnold H. Hara
461 W. Lanikaula St.
Univ. of Hawaii at Manoa
Hawaii Branch Station
Hilo, HI 96720
(808) 935-2885

Idaho

Hugh W. Homan
Dept. of Plant, Soil
  & Entomological Sciences
Univ. of Idaho
Moscow, ID 83843
(208) 885-7542

Illinois

Fredric D. Miller Jr.
1010 Jorie Blvd, Ste. 300
Univ. of Illinois
Oak Brook, IL 60521
(708) 920-0760

Phil Nixon
172 Natural Resources Bldg.
607 E. Peabody Dr.
Univ. of Illinois
Champaign, IL 61820
(217) 333-6650

Indiana

Timothy J. Gibb
Dept. of Entomology
Entomology Hall
Purdue Univ.
West Lafayette, IN 47907
(317) 494-4570

Clifford S. Sadof
Dept. of Entomology
Entomology Hall
Purdue Univ.
West Lafayette, IN 47907
(317) 494-5983

Iowa

Donald R. Lewis
Dept. of Entomology
Iowa State Univ.
Ames, IA 50011-3140
(515) 294-1101

Kansas

H. Leroy Brooks
Dept. of Entomology
Waters Hall
Kansas State Univ.
Manhattan, KS 66506
(913) 532-5891

Kentucky

Rudolph A. Scheibner
S225U Ag. Sci. Ctr.-No.
Univ. of Kentucky
Lexington, KY 40546
(606) 257-5838

Lee H. Townsend
S225M Ag. Sci. Ctr.-No.
Univ. of Kentucky
Lexington, KY 40546
(606) 257-7455

Louisiana

Dale K. Pollet
Knapp Hall
Louisiana State Univ.
Baton Rouge, LA 70803
(504) 388-4141

Maryland

J. Lee Hellman,
  John A. Davidson
Univ. of Maryland
College Park, MD 20742
(301) 454-3845

Michael J. Raupp
Univ. of Maryland
College Park, MD 20742
(301) 454-7025

Massachusetts

Robert Childs,
  Patricia J. Vittum
Fernald Hall
Univ. of Massachusetts
Amherst, MA 01003
(413) 545-2283

Maine

Clay Kirby
Pest Management Ofc.
491 College Ave.
Orono, ME 04469
(207) 581-3880

Michigan

David Smitley
Dept. of Entomology
Michigan State Univ.
East Lansing, MI 48824
(517) 355-3385

Minnesota

Mark E. Ascerno
236 Hodson Hall
Univ. of Minnesota
St. Paul, MN 55108
(612) 624-9272

Jeffrey D. Hahn
236 Hodson Hall
Univ. of Minnesota
St. Paul, MN 55108
(612) 624-4977

Mississippi

James H. Jarratt
Box 5446
Mississippi State Univ.
Mississippi State, MS 39762
(601) 325-3370

Missouri

Bruce A. Barrett
1-87 Agriculture Bldg.
Univ. of Missouri
Columbia, MO 65211
(314) 882-7894

Montana

Gary L. Jensen
Entomology Research Lab
Montana State Univ.
Bozeman, MT 59717
(406) 994-5690

Nebraska

Frederick P. Baxendale,
  J. Ackland Jones
210 Plant Industry-E. Campus
Univ. of Nebraska
LIncoln, NE 68583-0816
(402) 472-2125

Nevada

Clyde Sorenson
Univ. of Nevada
Reno, NV 89557
(702) 784-4419

New Hampshire

Stanley R. Swier
17 Nesmith Hall
Univ. of New Hampshire
Durham, NH 03824
(603) 862-1159

New Jersey

Louis M. Vasvary
Rutgers Cooperative Ext.
Rutgers-The State Univ.
Box 231
New Brunswick, NJ 08903
(201) 932-9324

New Mexico

Charles R. Ward
Route 1, Box 121
Southeastern Branch Station
New Mexico State Univ.
Artesia, NM 88210
(505) 748-1228

New York

Warren T. Johnson
Dept. of Entomology
Comstock Hall
Cornell Univ.
Ithaca, NY 14853-0999
(607) 255-4426

John P. Sanderson
Dept. of Entomology
Comstock Hall
Cornell Univ.
Ithaca, NY 14853-0999
(607) 255-5419

North Carolina

James R. Baker,
  Rick L. Brandenburg
Box 7613
North Carolina State Univ.
Raleigh, NC 27695
(919) 737-2703

North Dakota

Dept. of Entomology
North Dakota State Univ.
Fargo, ND 58105
(701) 237-7581

Ohio

Richard K. Lindquist
OARDC
Ohio State Univ.
Wooster, OH 44691
(216) 263-3736

David Shetlar
1991 Kenny Rd.
Ohio State Univ.
Columbus, OH 43210-1090
(614) 292-5274

Oklahoma

Ken N. Pinkston
501 Life Science West
Oklahoma State Univ.
Stillwater, OK 74078
(405) 624-5532

Oregon

Jack DeAngelis
Dept. of Entomology
2046 Cordley Hall
Oregon State Univ.
Corvallis, OR 97331-2907
(503) 737-3152

Pennsylvania

Paul R. Heller
106 Patterson Bldg.
Pennsylvania State Univ.
University Park, PA 16802
(814) 865-4621

Puerto Rico

Dept. of Crop Protection
Univ. of Puerto Rico
Mayaquez, PR 00708
(809) 832-4040

Rhode Island

Steven R. Alm
Dept. of Plant Sciences
Univ. of Rhode Island
Kingston, RI 02881
(401) 792-5998

Richard A. Casagrande
Dept. of Plant Pathology-
  Entomology
Univ. of Rhode Island
Kingston, RI 02881
(401) 792-2924

South Carolina

C. S. Gorsuch
111 Long Hall
Clemson Univ.
Clemson, SC 29634-0365
(803) 656-5043

South Dakota

Murdick McLeod
Dept. of Plant Science
South Dakota State Univ.
Brookings, SD 57007
(605) 688-5125

Tennessee

Harry E. Williams
Entomology and Plant
  Pathology Section
Univ. of Tennessee
Box 1071
Knoxville, TN 37901
(615) 974-7138

Texas

Philip J. Hamman
Entomology Dept.
Texas Λ & M Univ.
College Station, TX 77843
(409) 845-7026

Michael Merchant
17360 Coit Rd.
Texas A & M Univ.
Dallas, TX 75252
(214) 231-5362

Utah

Diane Alston
Dept. of Biology
UMC 5305
Utah State Univ.
Logan, UT 84322
(801) 750-2516

Vermont

Gordon R. Nielsen
Plant and Soil Science Dept.
Box 2164
Univ. of Vermont
S. Burlington, VT 05405-
  0082
(802) 656-2630

Virgin Islands

Jozef Keulants
Cooperative Ext. Service
RR 2, Box 10000
Kingshill, St. Croix 00850
(809) 778-0246

Virginia

Peter Schultz
Hampton Roads Agricultural
  Exp. Station
1444 Diamond Springs Rd.
Virginia Beach, VA 23455
(804) 363-3905

Washington

Arthur L. Antonelli
Western Washington
  Research & Ext. Ctr.
Washington State Univ.
Puyallup, WA 98371
(206) 840-4545

Elizabeth H. Beers
Tree Fruit Research Ctr.
Washington State Univ.
Wenatchee, WA 98801
(509) 663-8181

West Virginia

Linda Butler
Div. of Plant & Soil Sciences
West Virginia Univ.
Morgantown, WV 26506
(304) 293-6023

Wisconsin

Charles F. Koval
246-B Russell Labs
Univ. of Wisconsin
Madison, WI 53706
(608) 262-6429

CANADA

British Columbia

R.A. Costello
Ministry of Ag. and Food
17720 57th Avenue
Cloverdale, BC V3S 4P9
(604) 576-2911

Alberta

M. Y. Steiner
Alberta Environmental Centre
Bag 4000
Vegreville, AB T0B 4L0

Ontario

Graeme Murphy
Ministry of Ag. and Food
Vineland Station
Vineland, ON L0R 2E0
(416) 562-4147

A. B. Broadbent
Agriculture Canada
Research Station
Vineland Station, ON L0R
2E0
(416) 562-4113

Gillian Ferguson
Ministry of Agriculture and
Food
Research Station
Harrow, ON N0R 1G0
(519) 738-2251

Nova Scotia

Beverly MacPhail
Nova Scotia Dept. of Ag.
Kentville, NS B4N 1J5

B. M. Toms
Nova Scotia Dept. of Ag.
Box 550
Truro, NS B2N 5E3

Quebec

Andre Carrier
Ministry of Ag. Fisheries
and Food
200a, Chemin Ste-Foy (2e)
Quebec, QC G1R 4X6
(418) 646-9681

**U.S. EXTENSION FLORI-
CULTURE/ORNAMEN-
TAL HORTICULTURE
SPECIALISTS**

Alabama

Richard Shumack
Auburn Univ.
Alabama Coop. Ext. Service
116 Extension Hall
Auburn, AL 36849
(205) 826-4985

Alaska

Wayne Vandre
2221 E. Northern Lights Blvd.
No. 240
Anchorage, AK 99508
(907) 279-5582

Arizona

Brooks Taylor
Univ. of Arizona
Dept. of Plant Sciences
Tucson, AZ 85721
(602) 621-1945

Arkansas

Gerald Klingaman
Univ. of Arkansas
Dept. of Hort. and Forestry
Plant Science 314
Fayetteville, AR 72701
(501) 521-2603

California

Richard Evans
Univ. of California, Davis
Dept. of Hort.
Davis, CA 95616
(916) 752-6617

Colorado

Chi Won Lee
Colorado State Univ.
Dept. of Horticulture
Fort Collins, CO 80523
(303) 491-7119

Connecticut

Richard McAvoy
Univ. of Connecticut
Dept. of Plant Science
1376 Storrs Rd.
Storrs, CT 06268
(203) 486-3435

Delaware

David Tatnall
Univ. of Delaware
Dept. of Hort.
Townsend Hall
Newark, DE 19717-1303
(302) 451-2506

District of Columbia

Pamela Marshall
1351 Nicholson St. NW
Washington, DC 20011
(202) 282-7410

Florida

Benny Tjia
Univ. of Florida
Dept. of Ornamental Hort.
IFAS
1545 Fifield
Gainesville, FL 32611
(904) 392-7935

Georgia

Jeff Lewis
Univ. of Georgia
Dept. of Hort.
Athens, GA 30602
(404) 542-2340

Hawaii

Kenneth Leonherdt
Univ. of Hawaii
Dept. of Hort.
Manoa
3190 Maile Way
Honolulu, HI 96822
(808) 948-8909

Idaho

Bob Tripepi
Univ. of Idaho
Dept. of PSES
Moscow, ID 83843
(208) 885-6276

Illinois

Marvin Carbonneau
Univ. of Illinois
Dept. of Hort.
1013 Plant Sciences Lab
1201 S. Dorner
Urbana, IL 61801
(217) 333-2124

Indiana

Allen Hammer
Purdue Univ.
Dept. of Horticulture
West Lafayette, IN 47907
(317) 494-1335

Iowa

Jeff Iles
Iowa State Univ.
Dept. of Hort.
105 Horticulture Bldg.
Ames, IA 50010
(515) 294-0029

Kansas

Larry Leuthold
Kansas State Univ.
Umberger Hall
Manhattan, KS 66506
(913) 532-5820

Kentucky

Robert Anderson
Univ. of Kentucky
Dept. of Hort.
 & Arch. Landscape
N318 Agr. Science Center N.
Lexington, KY 40546
(606) 257-4721

Louisiana

Tom Pope
Louisiana State Univ.
Dept. of Hort.
Knapp Hall
Baton Rouge, LA 70803
(504) 388-2222

Maine

Lois Stack
Univ. of Maine
Dept. of Hort.
119 Deering Hall
Orono, ME 04469
(207) 581-2949

Maryland

William Healy
Univ. of Maryland
Dept. of Hort.
2107 Holzapfel Hall
University Park, MD 20742
(301) 454-8924

Massachusetts

Cathleen Carroll
Univ. of Massachusetts
Dept. of Plant & Soil Science
French Hall
Amherst, MA 01003
(413) 545-0895

Michigan

William Carlson
Michigan State Univ.
A240-E Plant & Soil Sciences
East Lansing, MI 48824
(517) 355-5178

Minnesota

John Erwin
Univ. of Minnesota
Dept. of Hort. Sciences
 and Landscape Architecture
St. Paul, MN 55108
(612) 624-9703

Mississippi

Richard Mullenax
Mississippi State Univ.
Dept. of Horticulture
P.O. Drawer T
Mississippi State, MS 39762
(601) 325-3223

Missouri

Univ. of Missouri-Columbia
Dept. of Horticulture
1-40 Agriculture Bldg.
Columbia, MO 65211
(314) 882-7511

Montana

George Evans
Plant and Soil Science
Leon Johnson Hall
Bozeman, MT 59717
(406) 994-4601

Nebraska

Jay Fitzgerald
Univ. of Nebraska
Dept. of Hort.
377L Plant Science Bldg.
Lincoln, NE 68538
(402) 472-1145

Nevada

Wayne Johnson
Univ. of Nevada
Dept. of Plant Science
Reno, NV 89557-0107
(702) 784-6911

New Hampshire

Charles Williams
Univ. of New Hampshire
Dept. of Hort.
Nesmith Hall
Durham, NH 03824
(603) 862-3207

New Jersey

George Wulster
Rutgers Univ.
Dept. of Hort. & Forestry
P.O. Box 231
Blake Hall
Cook College
New Brunswick, NJ 08903
(201) 932-8424

New Mexico

Lynn-Ellen Doxon
New Mexico State Univ.
9301 Indian School Rd., NE
Ste. 101
Albuquerque, NM 87112
(505) 275-5231

New York

Thomas Weiler
Cornell Univ.
Dept. of Floriculture
 and Ornamental Hort.
Ithaca, NY 14853
(607) 255-2166

North Carolina

Joseph Love
North Carolina State Univ.
Dept. of Hort. Science
Kilgore Hall
Raleigh, NC 27695
(919) 737-3322

North Dakota

Ron Smith
North Dakota State Univ.
Dept. of Hort.
Box 5658
Univ. Station
Fargo, ND 58105
(701) 237-8161

Ohio

Harry Tayama
Ohio State Univ.
Dept. of Hort.
241 Howlett Hall
2001 Fyffe Court
Columbus, OH 43210
(614) 292-9784

Oklahoma

David Buchanan
Oklahoma State Univ.
Dept. of Hort.
Stillwater, OK 74078
(405) 624-5414

Oregon

James Green
Oregon State Univ.
Dept. of Hort.
Corvallis, OR 97331
(503) 754-3464

Pennsylvania

Dennis Wolnick
Pennsylvania State Univ.
Dept. of Hort.
102 Tyson Bldg.
University Park, PA 16802
(814) 865-4040

Puerto Rico

Leornardo Flores
Univ. of Puerto Rico
Dept. of Hort.
Mayague Campus
Puerto Rico 00708
(809) 832-4040

Rhode Island

Richard Shaw
Univ. of Rhode Island
Dept. of Plant Sciences
Kingston, RI 02881
(401) 792-2791

South Carolina

John Kelly
Clemson Univ.
Dept. of Hort.
Clemson, SC 29634
(803) 656-2603

Tennessee

Douglas Crater
Univ. of Tennessee
Dept. of Ornamental Hort.
Box 1071
Knoxville, TN 37901
(615) 974-7324

Texas

Don Wilkerson
Texas A & M Univ.
Dept. of Hort.
College Station, TX 77843
(409) 845-5341

Utah

William Varga
Utah State Univ.
Dept. of Hort.
Logan, UT 84322-4820
(801) 750-2258

Vermont

Leonard Perry
Univ. of Vermont
Plant & Soil Science Dept.
Hills Bldg.
Burlington, VT 05405
(802) 656-2630

Virginia

Kevin Gruber
Virginia Polytechnic Inst.
  and State Univ.
Dept. of Hort.
301 Saunders Hall
Blacksburg, VA 24061
(703) 961-5451

Bonnie Appleton
Virginia Tech
Ext. Div.
Hampton Roads
Ag. Experimental Station
1444 Diamond Springs Rd.
Virginia Beach, VA 23455
(804) 446-4906

Washington

Paul Rasmussen
Washington State Univ.
Dept. of Hort.
Pullman, WA 99164-6414
(509) 335-9502

West Virginia

Richard Zimmerman
Cooperative Ext. Service
West Virginia Univ.
P.O. Box 6031
817 Knapp Hall
Morgantown, WV 26506-6031
(304) 293-5691

Wisconsin

Edward Hasselkus
Univ. of Wisconsin
Dept. of Hort.
1575 Linden Dr.
Madison, WI 53706
(608) 262-1450

Wyoming

James Cook
Univ. of Wyoming
Box 3354
University Station
Laramie, WY 82071
(307) 766-2243

PESTICIDE CONTROL OFFICIALS/CHEMISTRY LABS

Alabama

John Jinks, Dir.
State Chemical Lab
AL Dept. of Ag. and Ind.
P.O. Box 329
Auburn, AL 36830
(205) 888-3491

Arizona

Dwight Harder, Lab Mgr.
State Ag. Lab
P.O. Box 1586
Mesa, AZ 85211-1586
(602) 834-7152

Arkansas

Elvira Thompson,
  Chief Chemist
AR State Plant Board
P.O. Box 1069
Little Rock, AR 72203
(501) 225-1598

California

Willliam Cusick, Chief
Chemistry Laboratory Svcs.
Div. of Inspection Svcs.
CA Dept. of Food & Ag.
3292 Meadowview Rd.
Sacramento, CA 95832
(916) 427-4595

Colorado

Chemist, Standards Lab
Div. of Plant Ind.
CO Dept. of Ag.
700 Kipling St. #4000
Denver, CO 80215
(303) 239-4145

Connecticut

Lester Hankin,
 Chief Chemist
CT Ag. Experiment Station
P.O. Box 1106
New Haven, CT 06504
(203) 789-7219

Delaware

Teresa Crenshaw,
 State Chemist
DE Dept. of Ag.
2320 S. DuPont Hwy.
Dover, DE 19901
(302) 739-4811

Florida

G. Marshall Gentry, Chief
Bureau of Pesticide Lab
Div. of Chemistry
FL Dept. of Ag. &
 Consumer Svcs.
3125 Conner Blvd.
Tallahassee, FL 32399-1650
(904) 488-9375

Georgia

John F. Williams, Dir.
Laboratories Div.
GA Dept. of Ag.
Capitol Square
Atlanta, GA 30334
(404) 656-3647

Hawaii

Barry Brennan, Acting Chmn.
Dept. of Ag. Biochemistry
College of Trop. Agr.
 & Human Resources
Univ. of Hawaii, Manoa
3050 Maile Way
Honolulu, HI 96822
(808) 956-7306

Illinois

Warren Goetsh,
 Superintendent
Bureau of Laboratories
IL Dept. of Ag.
State Fairgrounds
Springfield, IL 62794-9281
(217) 785-8217

Indiana

Rodney J. Noel, Lab Dir.
IN State Chemist Office
1154 Biochemistry
Purdue Univ.
West Lafayette, IN 49707
(317) 494-5900

Iowa

Roger Bishop, Supervisor
Pesticide Lab
IA Dept. of Ag.
Wallace Bldg.
Des Moines, IA 50319
(515) 281-5861

Kansas

Max Foster, Dir.
Div. of Laboratories
KS Dept. of Ag.
2524 West 6th St.
Topeka, KS 66606
(913) 296-3301

Kentucky

Ron Egnew, Dir.
KY Dept. of Ag.
Capitol Plaza Tower
Frankfort, KY 40601
(502) 564-7274

Louisiana

Hershel F. Morris Jr., Dir.
Ag. Chemistry
LA Dept. of Ag.
P.O. Box 25060
Baton Rouge, LA 70894-5060
(504) 342-5812

Maryland

Warren Bontoyan
State Chemist
State Chemist Section
Plant Industries &
 Resource Conserv.
MD Dept. of Ag.
Annapolis, MD 20742
(301) 841-2721

Massachusetts

John M. Clark, Dir.
Dept. of Entomology
Fernald Hall
Univ. of Massachusetts
Amherst, MA 01003
(413) 545-2283

Michigan

T.K. Wu, Dir.
Laboratory Div.
MI Dept. of Ag.
1615 South Harrison Rd.
East Lansing, MI 48823
(517) 373-6410

Minnesota

William Krueger, Dir.
Laboratory Svcs.
MN Dept. of Ag.
90 West Plato Blvd.
St. Paul, MN 55107
(612) 296-3273

Mississippi

Earl G. Alley
State Chemist
MS State Chemical Lab
P.O. Box CR
Mississippi State, MS 39762
(601) 325-3324

Missouri

James R. Lea, Supvr.
Bureau of Pesticide Control
Plant Industries Div.
MO Dept. of Ag.
P.O. Box 630
Jefferson City, MO 65102
(314) 751-2462

Montana

Laszlo Torma, Chief
Lab Bureau
Environmental Mgt. Div.
MT Dept. of Ag.
McCall Hall, Room 10
Montana State Univ.
Bozeman, MT 59717
(406) 994-3383

Nebraska

Thomas Jensen, Chief
Ag.Laboratories
3703 South 14th St.
Lincoln, NE 68502
(402) 471-2176

Nevada

Harlan Specht,
 Chief Chemist
NV Dept. of Ag.
P.O. Box 11100,
 350 Capitol Hill Ave.
Reno, NV 89510-1100
(702) 789-0180

New Jersey

Eric Rau
Bureau of Env. Labs
NJ Dept. of Env. Protection
401 E. State St. CN 411
Trenton, NJ 08625
(609) 530-4119

New Mexico

Rick Janecka, State Chemist
State Chemist Lab
NM Dept. of Ag.
P.O. Box 30005
Dept. 3189
Las Cruces, NM 88003
(505) 646-3318

New York

Charles Weiss, Sr. Chemist
Hale Creek Field Station
7235 Steele Ave. Ext., RD #2
Gloversville, NY 12078
(518) 773-7318

North Carolina

Joel M. Padmore, Lab Dir.
Food & Drug Protection Div.
NC Dept. of Ag.
4000 Reedy Creek Rd.
Raleigh, NC 27607
(919) 733-7366

North Dakota

James L. Pearson, Dir.
ND Dept. of Health
 & Consolidated Labs
P.O. Box 937
Bismarck, ND 58502
(701) 221-6140

Ohio

Wilbur Highley
Lab Supvr., Residue
Robert Risner
Lab Supvr., Formulation
Consumer Analytical Labs
OH Dept. of Ag.
8995 E. Main St.
Reynoldsburg, OH 43068
(614) 866-6361

Oklahoma

Sue Cannon, Dir.
Agriculture Lab
OK Dept. of Ag.
2800 North Lincoln Blvd.
Oklahoma City, OK 73105
(405) 521-3864

Oregon

Michael Wehr, Adm.
Lab Svcs. Div.
OR Dept. of Ag.
635 Capitol St. NE
Salem, OR 97310
(503) 378-3793

Pennsylvania

Howard Nields, Chief
Laboratories Div.
PA Dept. of Ag.
2301 N. Cameron St.
Harrisburg, PA 17110
(717) 772-3235

Puerto Rico

Arline Gonzale, Dir.
Analysis & Reg. of
 Ag. Materials
PR Dept. of Ag.
P.O. Box 10163
Santurce, PR 00908
(809) 796-1710

Rhode Island

Public Health Chemist
Div. of Labs
RI Dept. of Health
50 Orms St.
Providence, RI 02904
(401) 277-3998

South Carolina

Jim Colcolough, Chemist IV
Ag. Chemical Svcs.
254 P & AS Bldg.
Clemson Univ.
Clemson, SC 29634-0391
(803) 656-3172

South Dakota

Duane Matthees
Station Biochemistry
SD State Univ.
Brookings, SC 57007-1217
(605) 688-6171

Tennessee

Joe Majors
Lab Supervisor
Div. of Plant Industries
TN Dept. of Ag.
P.O. Box 40627
Melrose Station
Nashville, TN 37204
(615) 360-0130

Texas

Rick Perry,
 Commissioner of Ag.
Lab Svcs.
TX Dept. of Ag.
P.O. Box 12847
Austin, TX 78711
(409) 836-8853

Utah

Ahmad Salari, State Chemist
UT State Ag. Dept.
350 N. Redwood Rd.
Salt Lake City, UT 84116
(801) 533-4131

Vermont

John Jaworski, Supvr.
Div. of Laboratories
VT Dept. of Ag.
103 S. Main St.
Waterbury, VT 05671
(802) 241-3008

Virginia

Dave Lynn, Supervisor
Residue Lab
Nancy Wieser, Supervisor
Formulations Lab
Consolidated Lab Svcs.
1 N. 14th St.
Richmond, VA 23219
(804) 786-3776

Washington

Herman Moya, Chief
Chemical & Hop Lab
2017 S. First St.
Yakima, WA 98903
(509) 575-2759

F. Clarke Brown, Chief
Compliance Section
WA Dept. of Ag.
2015 S. First St.
Yakima, WA 98903
(509) 575-2746

West Virginia

Ray Barber, Lab Dir.
Regulatory and Inspection
Div.
WV Dept. of Ag.
1900 Kanawha Blvd. East
Charleston, WV 25305
(304) 348-2208

Wisconsin

Gerald R. Myrdal, Dir.
Bureau of Lab Svcs.
4702 University Ave.
Madison, WI 53705
(608) 267-3500

Wyoming

Ken McMillan, Dir.
WY Dept. Ag. Analytical
  Services
1174 Snowy Range Rd.
Laramie, WY 82070
(307) 742-2984

**STATE EXTENSION
CLINIC SPECIALISTS**

Alabama

Jacqueline Mullen
Ext. Program Assoc.
Plant Pathologist
Extension Hall
Auburn University, AL 36849
(205) 826-4940

Colorado

William M. Brown
Ext. Plant Pathologist
Colorado State Univ.
Fort Collins, CO 80523
(303) 491-6470

District of Columbia

M.S. Khan
Ext. Pathologist/
  Pesticide Coord.
Plant Diagnostic Lab
Univ. of the D.C.
901 Newton St. NE
Washington, D.C. 20017
(202) 576-7419

Florida

Richard Cullen
  Biologist
Thomas A. Kucharek
  Prof. & Ext. Plant
  Pathologist
Gary W. Simone
  Assoc. Prof. & Ext.
  Plant Pathologist
Plant Pathology Dept.
Fifield Hall
Univ. of Florida
Gainesville, FL 32611
(904) 392-1980

Hawaii

Albert P. Martinez
Specialist, Plant Pathology
3190 Maile Way
Univ. of Hawaii
Honolulu, HI 96822
(808) 948-8053

Wayne T. Nishijima
Assoc. Specialist
Plant Pathology
875 Komohana St.
Hilo, HI 96720
(808) 959-9528

Illinois

Nancy Pataky
Plant Clinic
Univ. of Illinois
Urbana, IL 61801

Malcolm C. Shurtleff
Professor
N-533 Turner Hall
1102 S. Goodwin Avenue
Univ. of Illinois
Urbana, IL 61801
(217) 333-2478

Indiana

Gail Evans-Ruhl
Plant Diagnostic Clinic
Botany & Plant Pathology
Purdue Univ.
Lafayette, IN 47907
(317) 494-4641

Iowa

Paula Flynn
Iowa State Univ.
105 Bessey Hall
Ames, IA 50011
(515) 294-0581

Kansas

Judy O'Mara
Dept. of Plant Pathology
Throckmorton Hall, Rm 414
Kansas State Univ.
Manhattan, KS 66505
(913) 532-5810

Kentucky

Paul Bachi
Plant Diagnostician
Research & Ext. Ctr.
P.O. Box 469
Princeton, KY 42445
(502) 365-7541

Cheryl Kaiser
Plant Diagnostician
Dept. of Plant Pathology
Univ. of Kentucky
Lexington, KY 40546
(606) 257-8949

Maryland

Ethel M. Dutky
Faculty Ext. Asst.
Plant Diagnostic Clinic
Univ. of Maryland
College Park, MD 20742
(301) 454-3816

Massachusetts

D. R. Cooley
Dept. of Plant Pathology
Univ. of Massachusetts
Fernald Hall
Amherst, MA 01003
(413) 545-0179

R. W. Wick
Suburban Exp. Station
240 Beaver Street
Waltham, MA 02154
(617) 891-0650

Michigan

David L. Roberts
Disease Diagnostic Clinic
Dept. of Botany &
  Plant Pathology
Michigan State Univ.
East Lansing, MI 48824
(517) 355-4536

Minnesota

Jill Pokorny
Plant Disease Clinic/Dial U
Dept. of Plant Pathology
Univ. of Minnesota
St. Paul, MN 55108
(612) 625-6290

Mississippi

M.V. Patel
Ext. Plant Pathologist
P.O. Box 5426
Mississippi State Univ.
Mississippi State, MS 39762
(601) 325-3370

Missouri

Barbara Corwin
Dept. of Plant Pathology
Agriculture Bldg., Rm. 45
Univ. of Missouri
Columbia, MO 65211
(314) 882-3018

Einar W. Palm
Prof. & Ext. Plant
  Pathology Specialist
3-22 Agriculture Bldg.
Univ. of Missouri
Columbia, MO 65211
(314) 882-3018

Nebraska

Luanne V. Coziahr
Ext. Assistant
Dept. of Plant Pathology
Univ. of Nebraska
Lincoln, NE 68583-0722
(402) 472-2559

Nevada

John E. Maxfield
Assoc. Prof. & Ext. Plant
  Pathologist
Dept. of Plant Science
Univ. of Nevada
Reno, NV 89507
(702) 784-6911

New Jersey

Bruce B. Clarke
Ext. Specialist, Plant
  Pathology
Dept. of Plant Pathology
P.O. Box 231
Rutgers Univ.
New Brunswick, NJ 08903
(201) 932-9400

New York

Juliet Carroll
Ext. Assoc.
Dept. of Plant Pathology
Cornell Univ.
Ithaca, NY 14853
(607) 255-7850

North Carolina

R.K. Jones
Professor & Ext. Plant
  Pathology Specialist
Box 7616
North Carolina State Univ.
Raleigh, NC 27695-7616
(919) 737-3619

North Dakota

Martin Draper
Dept. of Plant Pathology
Box 5012
North Dakota State Univ.
Fargo, ND 58105
(701) 237-7854

Ohio

Stephen T. Nameth
Assoc. Prof. Plant Pathology
2021 Coffey Rd.
Ohio State Univ.
Columbus, OH 43210
(614) 292-6397

Oklahoma

Janet L. Jacobs
110 Noble Resident Center
Oklahoma State Univ.
Stillwater, OK 74078
(405) 624-5643

Oregon

Jeff Britt
Plant Clinic Supvr.
Cordley Hall
Oregon State Univ.
Corvallis, OR 97331
(503) 754-3472

Pennsylvania

John D. Peplinski
Coord., Plant Disease Clinic
218 Buckhout Laboratory
Pennsylvania State Univ.
University Park, PA 16802
(814) 865-1847

Tennessee

Steven C. Bost
  Assoc. Prof.
Elizabeth A. Long
  Ext. Assistant
Alan S. Windham
  Asst. Prof.
Ag. Ext. Service
Univ. of Tennessee
P.O. Box 110019
Nashville, TN 37211-0019
(615) 832-6802

M.A. Newman, Professor
Ag. Ext. Service
605 Airways Blvd.
Jackson, TN 38301
(901) 422-1583

Texas

Larry W. Barnes
Ext. Plant Pathologist
Room 101, L.F. Peterson
  Bldg.
Texas A & M Univ.
College Station, TX 77843
(409) 845-8032

Virginia

Mary Ann Hanson
Clinic Manager
Dept. of Plant Pathology,
  Physiology & Weed Science
VPI & SU
Blacksburg, VA 24061
(703) 961-6758

West Virginia

John F. Baniecki
Ext. Specialist,
  Plant Path. & Entomology
Safe Use of Pesticides &
  Chemicals
414 Brooks Hall, Main
  Campus
West Virginia State Univ.
Morgantown, WV 26502

Wisconsin

Craig R. Grau
Asst. Prof.
Dept. of Plant Pathology
1630 Linden Dr.
Univ. of Wisconsin
Madison, WI 53706
(608) 262-6289

**ORNAMENTAL PLANT
PATHOLOGISTS**

Alabama

Austin Hagan
Ext. Plant Pathologist
  and Nematologist
Extension Hall
Auburn Univ.
Auburn, AL 36849
(205) 826-4940

Alaska

Wayne Vandre
Ext. Horticulturist
2221 E. Northern Lights Blvd.
Anchorage, AK 99508-4143
(907) 279-6575

Arizona

Richard B. Hine, Professor
Dept. of Plant Pathology
Univ. of Arizona
Tucson, AZ 85721
(602) 621-1828

Arkansas

Gary L. Cloud
Ext. Plant Pathologist
P.O. Box 391
Univ. of Arkansas
Little Rock, AR 72203
(501) 671-2000

California

Arthur H. McCain
Ext. Plant Pathologist
Dept. of Plant Pathology
147 Hilgard Hall
Univ. of California
Berkeley, CA 94720
(415) 642-7153

Colorado

William M. Brown
Ext. Plant Pathologist
Colorado State Univ.
Fort Collins, CO 80523
(303) 491-6470

Connecticut

David B. Schroeder
Prof. & Ext. Plant Pathologist
Natural Resources
  Conservation
Univ. of Connecticut
Storrs, CT 06268
(203) 486-2839

Delaware

Robert P. Mulrooney
Ext. Plant Pathologist
Townsend Hall, R-154
Univ. of Delaware
Newark, DE 19717-1303
(302) 451-2534

District of Columbia

M.S. Khan
Ext. Pathologist/
  Pesticide Coordinator
Plant Diagnostic Lab.
Univ. of the D.C.
901 Newton St. NE
Washington, DC 20017
(202) 576-7419

Florida

Gary W. Simone
Assoc. Prof. and
  Ext. Plant Pathologist
Fifield Hall
Plant Pathology Dept.
Univ. of Florida
Gainesville, FL 32611
(904) 392-1980

Georgia

Eugene H. Moody
Ext. Plant Pathologist
Cooperative Ext. Service
The Univ. of Georgia
Athens, GA 30602
(404) 542-2685

Guam

W.P. Leon Guerrero
Dir., Coop. Ext. Service
Univ. of Guam
Box EK
Agana, Guam 96910
(671) 734-9162

Hawaii

Albert P. Martinez
Specialist, Plant Pathology
3190 Maile Way
Univ. of Hawaii
Honolulu, HI 96822
(808) 948-8053

Idaho

Robert Forster
Ext. Plant Pathologist
Univ. of Idaho
Research & Extension Ctr.
Kimberly, ID 83341
(208) 423-4691

Illinois

Malcolm C. Shurtleff, Prof.
N-533 Turner Hall
1102 S. Goodwin Ave.
Univ. of Illinois
Urbana, IL 61801
(217) 333-2478

Indiana

Paul Pecknold
Assoc. Prof.
Botany and Plant Pathology
Purdue Univ.
Lafayette, IN 47907
(317) 494-4628

Iowa

Mark Gleason
Asst. Prof.
Dept. of Plant Pathology,
  Seed & Weed Science
Iowa State Univ.
Ames, IA 50011
(515) 294-1160

Kansas

Ned A. Tisserat
Ext. Plant Pathologist
Dept. of Plant Pathology
Throckmorton Hall
Kansas State Univ.
Manhattan, KS 66505
(913) 532-5810

Kentucky

John R. Hartman
Ext. Prof.
Dept. of Plant Pathology
Univ. of Kentucky
Lexington, KY 40546
(606) 257-5779

Louisiana

Kenneth Whitam
Specialist & Project Leader
Plant Pathology
250 Knapp Hall
Louisiana State Univ.
Baton Rouge, LA 70803
(504) 388-4562

Maine

Steve Johnson
Ext. Plant Pathologist
  & Potato Specialist
Aroostook Co. Ext. Office
Houlton Rd.
Presque Isle, ME 04769
(207) 581-3311

Maryland

Paul W. Steiner
Assoc. Prof. & Ext.
  Plant Pathologist
Univ. of Maryland
College Park, MD 20742
(301) 454-3816

Massachusetts

R.W. Wick
Suburban Experiment Station
240 Beaver Street
Waltham, MA 02154
(617) 891-0650

Michigan

Christine Taylor-Stephens
Asst. Prof.
Dept. of Botany and
  Plant Pathology
140 Plant Biology Bldg.
Michigan State Univ.
East Lansing, MI 48824-1312
(517) 355-4534

Minnesota

Francis L. Pfleger
Ext. Plant Pathologist
Dept. of Plant Pathology
University of Minnesota
St. Paul, MN 55108
(612) 625-6290

Mississippi

Robert A. Haygood
Ext. Plant Pathologist
P.O. Box 5446
Mississippi State Univ.
Mississippi State, MS 39762
(601) 325-3370

Missouri

Einar W. Palm
Prof. Plant Pathology
  & State Ext. Plant
  Pathology Specialist
3-22 Agriculture Bldg.
Univ. of Missouri
Columbia, MO 65211
(314) 882-3018

Montana

Jack Riesselman
Ext. Plant Pathologist
Dept. of Plant Pathology
525 Johnson Hall
Montana State Univ.
Bozeman, MT 59717
(406) 994-4832

Nebraska

John E. Watkins
Prof. & Ext. Plant Pathologist
Dept. of Plant Pathology
Univ. of Nebraska
Lincoln, NE 68583-0722
(402) 472-2559

Nevada

John E. Maxfield
Assoc. Prof. & Ext.
  Plant Pathologist
Dept. of Plant Science
Univ. of Nevada
Reno, NV 89557-0004
(702) 784-6911

New Hampshire

William E. MacHardy
Ext. Plant Pathologist
325 Nesmith Hall
Univ. of New Hampshire
Durham, NH 03824
(603) 862-2060

New Jersey

Bruce B. Clarke
Ext. Specialist,
  Plant Pathology
Dept. of Plant Pathology
P.O. Box 231
Rutgers Univ.
New Brunswick, NJ 08903
(908) 932-9400

John K. Springer
Ext. Specialist,
  Plant Pathology
Rutgers Research &
  Development
RR 5
Bridgeton, NJ 08302
(609) 455-3100

New Mexico

Emroy L. Shannon
Ext. Plant Pathologist
Box 3AE
New Mexico State Univ.
Las Cruces, NM 88003
(505) 646-1822

New York

Margery L. Daughtrey
Ext. Assoc.
Cornell University
Horticulture Research Lab
Riverhead, NY 11901
(516) 727-3595

R.K. Horst, Professor
Dept. of Plant Pathology
Cornell Univ.
Ithaca, NY 14853
(607) 255-7859

North Carolina

R.K. Jones
Prof. & Ext. Plant Pathology
  Specialist
Box 7616
North Carolina State Univ.
Raleigh, NC 27695-7616
(919) 737-3619

North Dakota

H. Arthur Lamey
Ext. Plant Pathologist
Box 5012
North Dakota State Univ.
Fargo, ND 58105
(701) 237-7056

Ohio

Charles C. Powell Jr.
Ext. Specialist & Prof. of
  Plant Pathology
2021 Coffey Rd.
Ohio State Univ.
Columbus, OH 43210
(614) 292-6397

Oklahoma

Larry J. Littlefield, Head
Dept. of Plant Pathology
Life Sciences East
Oklahoma State Univ.
Stillwater, OK 74078
(405) 624-5643

Oregon

Iain C. MacSwan
Plant Pathology Specialist
Cordley Hall
Oregon State Univ.
Corvallis, OR 97331
(503) 754-3472

Pennsylvania

Gary Moorman
Asst. Prof.
211 Buckhout Laboratory
Pennsylvania State Univ.
University Park, PA 16802
(814) 865-1847

Puerto Rico

Lii-Jang Liu, Dir.
Crop Protection
Univ. of Puerto Rico
Mayaguez, PR 00708
(809) 832-4040

Rhode Island

David B. Wallace
Ext. Specialist
Plant Protection
Dept. of Plant Sciences
Univ. of Rhode Island
Kingston, RI 02881
(401) 792-2481

South Carolina

R.W. Miller
Prof. of Plant Pathology
  and Physiology
206 Long Hall
Clemson Univ.
Clemson, SC 29634
(803) 656-5732

South Dakota

Fred Cholick
Dept. Head
Plant Science Dept.
South Dakota State Univ.
Brookings, SD 57007
(605) 688-5121

Tennessee

Alan S. Windham
Asst. Prof.
Ag. Ext. Service
Univ. of Tennessee
P.O. Box 110019
Nashville, TN 37222-0019
(615) 832-6802

Texas

Jerrel D. Johnson
Ext. Plant Pathologist
Room 118, L.F. Peterson
 Bldg.
Texas A & M Univ.
College Station, TX 77843
(409) 845-8032

Utah

Sherman V. Thomson
Ext. Plant Pathologist
Dept. of Biology
Utah State Univ.
Logan, UT 84322-5305
(801) 750-3406

Vermont

Alan R. Gotlieb
Ext. Plant Pathologist
Dept. of Plant & Soil Science
Univ. of Vermont
Hills Bldg.
Burlington, VT 05405
(802) 656-2630

Virgin Islands

Walter Knausenberger
Pest Mgt. Specialist
Coop. Ext. Service
College of Virgin Islands
P.O. Box L, Kingshill
St. Croix, VI 00850
(809) 778-0246

Virginia

R. Jay Stipes
Ext. Specialist,
 Plant Pathology
Dept. of Plant Pathology,
 Physiology & Weed Science
VPI & SU
Blacksburg, VA 24061
(703) 961-7479

Washington

Ralph S. Byther
Ext. Plant Pathologist
Western Washington
 Research & Ext. Ctr.
Puyallup, WA 98371
(206) 845-6613

Otis C. Maloy
Ext. Plant Pathologist
Washington State Univ.
Pullman, WA 99164
(509) 335-9541

West Virginia

John F. Baniecki
Ext. Specialist, Plant
 Pathology & Entomology
Safe Use of Pesticides &
 Chemicals
414 Brooks Hall, Main
 Campus
West Virginia State Univ.
Morgantown, WV 26506
(304) 293-3911

Wisconsin

Gayle L. Worf, Professor
Dept. of Plant Pathology
1630 Linden Dr.
Univ. of Wisconsin
Madison, WI 53706
(608) 262-1410

Wyoming

Don A. Roth
Ext. Plant Pathologist
P.O. Box 3354
University Station
Laramie, WY 82071
(307) 766-2248

REGIONAL POISON CONTROL CENTERS

Alabama

Alabama Poison Ctr.
809 University Blvd. E.
Tuscaloosa, AL 35401
(205) 345-0600

Arizona

Arizona Poison
 Control System
Arizona Health Sciences Ctr.
1501 N. Campbell
Rm. 3204 K
Univ. of Arizona
Tucson, AZ 85724
(602) 626-6016

California

California Poison Centers
Los Angeles Co. Medical
 Assn.
Regional Poison Control Ctr.
1925 Wilshire Blvd.
Los Angeles CA 90057
(213) 484-5151

San Diego Regional
 Poison Ctr.
UCSD Medical Center
225 Dickinson St.
San Diego, CA 92103
(619) 543-6000

San Francisco Bay Area
 Regional Poison Control Ctr.
San Francisco General
 Hospital
Rm. 1E86
1001 Potrero Ave.
San Francisco, CA 94110
(415) 476-6600

UCDMC Regional
 Poison Control Ctr.
2315 Stockton Blvd.
Sacramento, CA 95817
(916) 734-3692

Colorado

Colorado Rocky Mountain
 Poison Ctr.
645 Bannock St.
Denver, CO 80204-4507
(303) 893-7774

District of Columbia

DC National
 Capital Poison Ctr.
Georgetown Univ. Hospital
3800 Reservoir Rd. NW
Washington, DC 20007
(202) 625-3333

Florida

Tampa Bay Regional
 Poison Control Ctr.
P.O. Box 1289
Tampa, FL 33601
(813) 253-4444

Georgia

Georgia Poison Control Ctr.
Box 26066
80 Butler St. SE
Atlanta, GA 30335
(404) 589-4400

Indiana

Indiana Poison Ctr.
Methodist Hospital
1701 N. Senate Blvd.
Indianapolis, IN 46206
(317) 929-2323

Kentucky

Kentucky Regional
  Poison Ctr.
Kosair Children's Hospital
P.O. Box 35070
Louisville, KY 40232-5070
(502) 589-8222

Louisiana

Louisiana Regional
  Poison Control Ctr.
Louisiana State Univ.
  School of Medicine
P.O. Box 33932
Shreveport, LA 71130-3932
(800) 535-0525

Maryland

Maryland Poison Ctr.
20 North Pine St.
Baltimore, MD 21201
(301) 528-7701

Massachusetts

Massachusetts Poison
  Control System
300 Longwood Ave.
Boston, MA 02115
(617) 232-2120

Michigan

Michigan Poison Ctr.
Blodgett Regional Poison Ctr.
1840 Wealthy SE
Grand Rapids, MI 49506
(616) 774-7444

Poison Control Ctr.
Children's Hospital of
  Michigan
3901 Beaubien Blvd.
Detroit, MI 48201
(313) 745-5711

Minnesota

Hennepin Regional
  Poison Ctr.
Hennepin Co. Medical Ctr.
701 Park Ave.
Minneapolis, MN 55415
(612) 347-3141

Minnesota Regional
  Poison Ctr.
St. Paul-Ramsey Medical Ctr.
640 Jackson St.
St. Paul, MN 55101
(612) 221-2113

Missouri

Cardinal Glennon
  Children's Hospital
Regional Poison Ctr.
1465 South Grand Blvd.
St. Louis, MO 63104
(314) 772-5200

Montana

Montana Poison Ctr.
645 Bannock St.
Denver, CO 80204-4507
(303) 893-7774

Nebraska

Nebraska Mid Plains
  Poison Ctr.
8301 Dodge St.
Omaha, NE 68114
(402) 390-5555

New Jersey

New Jersey Poison
  Information and
  Education System
201 Lyons Ave.
Newark, NJ 07112
(201) 926-8005

New Mexico

New Mexico Poison and
  Drug Information Ctr.
Univ. of New Mexico
Albuquerque, NM 87131
(505) 843-2551

New York

New York Poison Centers
Long Island Regional
  Poison Control Center
Nassau Co. Medical Center
2201 Hempstead Turnpike
East Meadow, NY 11554
(516) 542-2323

New York City
  Poison Control Ctr.
455 First Ave., Rm. 123
New York, NY 10016
(212) 340-4494

North Carolina

Duke Univ. Poison
  Control Ctr.
Box 3007
Duke Univ. Medical Ctr.
Durham, NC 27710
(919) 684-8111

Ohio

Central Ohio Poison Ctr.
Columbus Children's
  Hospital
700 Children's Dr.
Columbus, OH 43205
(800) 682-7625

Southwest Ohio Regional
  Poison Control System
Drug and Poison
  Information Ctr.
231 Bethesda Ave.,
  M.L. No. 144
Cincinnati, OH 45267-0144
(800) 872-5111

Pennsylvania

Pittsburgh Poison Ctr.
Children's Hospital
  Pittsburgh
3705 Fifth Ave. at DeSoto St.
Pittsburgh, PA 15213
(412) 681-6669

Rhode Island

Rhode Island Poison Ctr.
Rhode Island Hospital
593 Eddy St.
Providence, RI 02902
(401) 277-5727

Texas

North Central Texas
  Poison Ctr.
P.O. Box 35926
Dallas, TX 75235
(214) 590-5000

Texas State Poison Ctr.
Univ. of Texas Medical
  Branch
Galveston, TX 77550-2780
(409) 772-1420

Utah

Intermountain Regional
Poison Control Ctr.
50 N. Medical Dr.
Bldg. 428
Salt Lake City, UT 84132
(801) 581-7504

West Virginia

West Virginia Poison Ctr.
West Virginia Univ.
 School of Pharmacy
3110 MacCorkle Ave., SE
Charleston, WV 25304
(304) 293-0111

STATE ENVIRON-
MENTAL PROTECTION
AGENCY OFFICES

Alabama

Dept. of Environmental Mgt.
Land Div.
1751 Cong Dr.
Montgomery, AL 36130⁻
(205) 271-7737

Alaska

Dept. of Environmental
 Conservation
Hazardous Waste Program
Pouch O
Juneau, AK 99811
(907) 465-2666

Arizona

Arizona Dept. of Health Svcs.
1740 W. Adams
Phoenix, AZ 85007
(602) 542-1000

Arkansas

Arkansas Dept. of Pollution
 Control and Ecology
Solid and Hazardous Waste
Div.
P.O. Box 8913
Little Rock, AR 72219
(501) 562-7444

California

Dept. of Health Svcs.
714 P St.
Sacramento, CA 95814
(916) 324-1781

Colorado

Colorado Dept. of Health
Waste Management Div.
4210 East 11th Ave.
Denver, CO 80220
(303) 320-8333

Connecticut

Connecticut Dept. of
 Environmental Protection
Bureau of Waste Management
165 Capitol Ave.
Hartford, CT 06106
(203) 566-5712

Delaware

Delaware Dept. of Natural
 Resources and
 Environmental Control
Solid and Hazardous Waste
 Mgt. Branch
89 Kings Hwy., P.O.
 Box 1401
Dover, DE 19901
(302) 739-3672

District of Columbia

Dept. of Consumer and
 Regulatory Affairs
Pesticides and Hazardous
 Waste Branch
5010 Overlook Ave. SW
Room 114
Washington D.C. 20032
(202) 404-1167

Florida

Florida Dept. of
 Environmental Regulation
Solid and Hazardous Waste
2600 Blair Stone Rd.
Tallahassee, FL 32399-2400
(904) 488-0300

Georgia

Georgia Dept. of Natural
 Resources
Land Protection Branch
205 Butler St.
Floyd Tower E., Ste. 1154
Atlanta, GA 30334
(404) 656-2833

Guam

Guam EPA
P.O. Box 2999
Agana, Guam 96910
(Overseas operator) 646-2833

Hawaii

Hawaii Dept. of Health
Environmental Protection
 and Health Svcs. Div.
Noise and Radiation Branch
P.O. Box 3378
Honolulu, HI 96810
(808) 548-3075

Idaho

Dept. of Environmental
 Quality
Bureau of Hazardous Waste
 Section
1410 N. Hilton
Boise, ID 83706
(208) 334-5879

Illinois

Illinois Environmental
 Protection Agency
Div. of Land Pollution
 Control
2200 Churchill Rd.
Springfield, IL 62794
(217) 782-6761

Indiana

Indiana Dept. of
 Environmental Management
Hazardous Waste Mgt.
 Branch
105 S. Meridian St.
Indianapolis, IN 46207-7035
(317) 232-7959

Iowa

U.S. EPA Region 7
726 Minnesota Ave.
Kansas City, KS 66101
(913) 551-7000

Kansas

Kansas Dept. of Health
 and Environment
Waste Management Bureau
Forbes Field
Topeka, KS 66620
(913) 296-1593

Kentucky

Dept. of Environmental
 Protection
Div. of Waste Management
18 Reilly Rd.
Frankfort, KY 40601
(502) 564-6717

Louisiana

Louisiana Dept. of
 Environmental Quality
Office of Solid and
 Hazardous Waste
P.O. Box 82178
Baton Rouge, LA 70804
(504) 765-0249

Maine

Maine Dept. of
 Environmental Protection
Bureau of Oil and Hazardous
 Materials Control
State House, Station 17
Augusta, ME 04333
(207) 289-2651

Maryland

Office of Environmental
 Programs
Waste Mgt. Administration
2500 Browning Hwy.
Baltimore, MD 21224
(301) 631-3400

Massachusetts

Dept. of Environmental
 Protection
Div. of Solid & Hazardous
 Waste
One Winter St.
Boston, MA 02108
(617) 292-5960 or 5853

Michigan

Dept. of Natural Resources
Hazardous Waste Div.
P.O. Box 30241
Lansing, MI 48909
(517) 373-2730

Minnesota

Minnesota Pollution
 Control Agency
Solid Waste Div.
520 Lafayette Rd.
St. Paul, MN 55155
(612) 296-7340

Mississippi

Dept. of Environmental
 Quality
Office of Pollution Control
Div. of Solid Waste Mgt.
P.O. Box 10385
Jackson, MS 39289
(601) 961-5171

Missouri

Missouri Dept. of
 Natural Resources
Div. of Environmental
 Quality
P.O. Box 1368
Jefferson City, MO 65102
(314) 751-4810

Montana

Dept. of Health and
 Environmental Sciences
Solid and Hazardous Waste
 Management Bureau
Cogswell Bldg, Rm. B-201
Helena, MT 59620
(406) 444-2821

Nebraska

Dept. of Environmental
 Control
Hazardous Waste Mgt.
 Section
P.O. Box 98922
301 Centennial Mall S.
Lincoln, NE 68509
(402) 471-2186

Nevada

Dept. of Conservation
 & Natural Resources
Div. of Environmental
 Protection
Capitol Complex
123 W. Nye Lane
Carson City, NV 89710
(702) 687-4670

New Hampshire

Div. of Public Health Svcs.
Office of Waste Mgt.
Health and Welfare Bldg.
6 Hazen Dr.
Concord, NH 03301
(603) 271-2900

New Jersey

Dept. of Environmental
 Protection
Div. of Waste Mgt.
Hazardous Waste Advisory
 Program
P.O. Box CN028
401 E. State St.
Trenton, NJ 08625
(609) 292-9120

New Mexico

New Mexico Environmental
 Enforcement Div.
Hazardous Waste Section
P.O. Box 26110
Santa Fe, NM 87502
(505) 827-2929

New York

Dept. of Environmental
 Conservation
Div. of Solid & Hazardous
 Waste
Manifest Section
50 Wolf Rd.
Albany, NY 12233
(518) 457-0530

North Carolina

Dept. Environmental Health
 & Natural Resources
Management Branch
P.O. Box 27687
Raleigh, NC 27611
(919) 733-2178

North Dakota

North Dakota State
 Dept. of Health
Div. of Hazardous Waste
Management and Special
 Studies
P.O. Box 5520
Bismarck, ND 58502
(701) 224-2366

Ohio

Ohio Environmental
 Protection Agency
Div. of Solid and Hazardous
 Waste Management
1800 Watermark Dr.
Columbus, OH 43266-0149
(614) 466-7220

Oklahoma

Oklahoma State
 Dept. of Health
Waste Mgt. Section
100 N.E. 10th St.
Oklahoma City, OK 73117
(405) 271-5338

Oregon

Dept. of Environmental
  Quality
Hazardous & Solid Waste
  Div.
1811 S.W. 6th Ave.
Portland, OR 97204
(503) 229-5913

Pennsylvania

Dept. of Environmental
  Resources
Bureau of Solid Waste Mgt.
P.O. Box 2063
Harrisburg, PA 17105
(717) 787-6239

Puerto Rico

Environmental Quality Board
Land Pollution Control Area
P.O. Box 11488
Santurce, PR 00910-1488
(809) 722-0439

Rhode Island

Dept. of Environmental Mgt.
Div. of Air & Hazardous
  Waste Mgt.
204 Cannon Bldg.
291 Promenade St.
Providence, RI 02908
(401) 277-2797

South Carolina

Dept. of Health and
  Environmental Control
Bureau of Solid & Hazardous
  Waste Mgt.
RCRA, J. Marion Sims Bldg.
2600 Bull St.
Columbia, SC 29201
(803) 734-5000

South Dakota

Environmental and Natural
  Resources & Air Quality
  Ofc.
Rm. 217, Foss Bldg.
523 E. Capitol
Pierre, SD 57501
(605) 773-3329

Tennessee

Div. of Solid Waste Mgt.
Customs House
4th Floor
701 Broadway
Nashville, TN 37243
(615) 741-3424

Texas

Texas Dept. of
  Water Resources
Industrial & Industrial Service
  Facilities
Industrial Solid Waste Section
P.O. Box 13087
Capitol Station
Austin, TX 78711
(512) 463-7761

Texas Dept. of Health,
  Commercial Service
Municipal State and
  Federal Facilities
Bureau of Solid Waste Mgt.
1100 West 49th St.
Austin, TX 78756
(512) 458-7271

Utah

Dept. of Environmental
  Health & Solid and
  Hazardous Waste
288 North, 1460 West St.
State Office Bldg., Rm. 4321
Salt Lake City, UT 84144
(801) 538-6170

Vermont

Agency of Environmental
  Conservation
Air and Solid Waste Div.
103 S. Main St.
Waterbury, VT 05676
(802) 828-3395

Virginia

Dept. of Health
Div. of Solid & Hazardous
  Waste Mgt.
Monroe Bldg., 11th Floor
101 North 14th St.
Richmond, VA 23219
(804) 225-2667

Virgin Islands

Dept. of Conservation
  & Cultural Affairs
Hazardous Waste Program
Div. of Natural Resources
  Mgt.
P.O. Box 4340
Charlotte Amalie, St. Thomas
Virgin Islands 00801
(809) 774-3320

Washington

Dept. of Ecology
Ofc. of Hazardous Substances
Mail Stop PV-11
Olympia, WA 98504
(206) 459-6299

West Virginia

Dept. of Natural Resources
Solid & Hazardous Waste
  Ground Water Branch
Div. of Water Resources
1356 Hansford St.
Charleston, WV 25311
(304) 348-5935

Wisconsin

Dept. of Natural Resources
Bureau of Solid Waste Mgt.
P.O. Box 7921
Madison, WI 53707
(608) 266-1327

Wyoming

EPA Region 8
Waste Management Div.
Hazardous Waste Branch
  (8HWM-ON)
999 18th St.
Denver, CO 80202-2405
(303) 293-1502

DIRECTORY OF EQUIPMENT, SUPPLIES AND SERVICES

(These lists are provided for
information only. No endorsement is intended for suppliers/
manufacturers mentioned, nor
is criticism meant for suppli-
ers/manufacturers not mentioned.)

Absorbents

Aquatrols Corp. of America
1432 Union Ave.
Pennsauken, NJ 08110
(609) 665-1130

Broadleaf Industries
7014 Manya Circle
San Diego, CA 92154
(619) 424-7880

Gro-Prod Inc.
1078 Rt. 46 W.
Clifton, NJ 07013
(201) 471-1234

Industrial Svcs. Intl. Inc.
Box 10834
Bradenton, FL 34282-0834
(813) 792-7778

JRM Chemical Div.
110 W. Streetsboro St.
Hudson, OH 44236
(216) 656-4010

Miller Chemical &
  Fertilizer Corp.
Box 333
Hanover, PA 17331
(717) 632-8921

Moisture Mizer Div.
Multiple Concepts Inc.
Box 4248
Chattanooga, TN 37405
(615) 266-3967

Soil-Tec Inc.
Box 59413
Dallas, TX 75229
(214) 263-0142

Vermiculite Ltd. Inc.
1078 Rt. 46 W.
Clifton, NJ 07013
(201) 471-1234

Adhesives

Dorothy Biddle Service
Dept. G
Greeley, PA 18425-9799
(717) 226-3239

Century Rain Aid
31691 Dequindre
Madison Heights, MI 48071
(313) 588-2990

Floc-Flo Corp.
211 N. Carpenter St.
Chicago, IL 60607
(312) 666-7000

Hendrix & Dail Inc.
Box 648
Greenville, NC 27835-0648
(919) 758-4263

Miller Chemical &
  Fertilizer Corp.
Box 333
Hanover, PA 17331
(717) 632-8921

Reef Industries
  Griffolyn Div.
Box 750250
Houston, TX 77275-0250
(713) 943-0070

Truxes Co.
16 Stonehill Rd.
Oswego, IL 60543
(708) 554-8448

Adjuvants

Bonide Chemical
2 Wurz Ave.
Yorkville, NY 13495
(315) 736-8231

EnP Inc.
2001 N. Main St.
Box 218
Mendota, IL 61342
(815) 539-7471

Miller Chemical &
  Fertilizer Corp.
Box 333
Hanover, PA 17331
(717) 632-8921

National Research
  & Chemical
14439 S. Avalon Blvd.
Gardena, CA 90248
(213) 515-1700

Sierra Crop Protection
12101 Woodcrest
  Executive Way
St. Louis, MO 63141
(314) 275-7561

Valent U.S.A. Corp.
Box 8025, Room 6051
Walnut Creek, CA 94596
(415) 256-2725

Wilbur-Ellis Co.
Box 47907
San Antonio, TX 78265-7907
(512) 227-5255

**Alarm systems,
environmental**

Argus Supply Co.
Box 689
Roseville, MI 48066
(313) 774-8900

Automata Inc.
19393 Redberry Rd.
Grass Valley, CA 95945
(916) 273-0380

Ball Seed Co.
622 Town Rd.
West Chicago, IL 60185
(708) 231-3500

Godro Inc.
Box 1682
Bloomington, IL 61701-1682
(309) 829-4353

Metex Corp. Ltd.
12 Penn Dr., Unit 1
Weston, ON M9L 2A9
CANADA
(416) 749-1210

Midwest Trading,
  GROmaster Div.
Box 384
St. Charles, IL 60174
(708) 742-1840

Phonetics Inc.
101 State Rd.
Media, PA 19063
(215) 565-8520

Q-Com Inc.
2050 S. Grand Ave.
Santa Ana, CA 92705
(714) 540-6123

Wadsworth Control Systems
5541 Marshall St.
Arvada, CO 80002
(303) 424-4461

Winland Electronics Inc.
Box 473
Mankato, MN 56001
(507) 625-7231

Alkaline water buffers

National Research
  & Chemical
14439 S. Avalon Blvd.
Gardena, CA 90248
(213) 515-1700

Boilers, hot water

Ball Seed Co.
622 Town Rd.
West Chicago, IL 60185
(708) 231-3500

Bio-Energy Systems Inc.
Box 191
Ellenville, NY 12428
(914) 647-6700

Biotherm Engineering
Box 6007
Petaluma, CA 94953
(707) 763-4444

CanaBurn Bioenergy Inc.
Box 2037
Windsor, ON N8Y 4R5
CANADA
(519) 945-5244

Delta T Sales
3576 Empleo, Unit 2
San Luis Obispo, CA 93401
(805) 546-8814

Eshland Enterprises Inc.
Box 8A
Greencastle, PA 17225
(717) 597-3196

Firetender-Stokermatic
  Coal-Stoker Heating Eqt.
1610 S. Industrial Rd.
Salt Lake City, UT 84104
(801) 972-4488

Hamilton Engineering Inc.
34155 Industrial Rd.
Livonia, MI 48150
(313) 522-5530

Vary Industries
Box 160
Grimsby, ON L3M 4N6
CANADA
(416) 945-9691

Winandy Ghse. Co. Inc.
2211 Peacock Rd.
Richmond, IN 47374
(317) 935-2111

Boilers, maintenance supplies

Biotherm Engineering
Box 6007
Petaluma, CA 94953
(707) 763-4444

Delta T Sales
3576 Empleo, Unit 2
San Luis Obispo, CA 93401
(805) 546-8814

Firetender-Stokermatic
  Coal-Stoker Heating Eqt.
1610 S. Industrial Rd.
Salt Lake City, UT 84104
(801) 972-4488

Books, gardening

The John Henry Co.
Box 17099
Lansing, MI 48901
(517) 323-9000

Horta-Craft Ltd.
RR 1, Mallard Rd.
Hyde Park, ON N0M 1Z0
CANADA
(519) 472-5120

Books, greenhouse production

Ball Publishing
P.O. Box 532
Geneva, IL 60134
(708) 208-9080

Dansco Distributors
4442 27th Ave., W.
Seattle, WA 98199
(206) 282-7282

Ohio Florists' Assn.
2130 Stella Ct., Ste. 200
Columbus, OH 43215
(614) 487-1117

Professional Plant
  Growers Assn.
P.O. Box 27517
Lansing, MI 48909
(517) 694-7700

Books, horticultural reference

Ball Publishing
P.O. Box 532
Geneva, IL 60134
(708) 208-9080

Horta-Craft Ltd.
RR1, Mallard Rd.
Hyde Park, ON N0M 1Z0
CANADA
(519) 472-5120

Chemicals, algaecides

DelTek Inc.
Box 179
Pearland, TX 77588
(800) 367-3951

Hendrix & Dail Inc.
Box 648
Greenville, NC 27835-0648
(919) 758-4263

Lesco Inc.
20005 Lake Rd.
Rocky River, OH 44116
(216) 333-9250

Lebanon Chemical Corp.
1600 E. Cumberland St.
Lebanon, PA 17042
(717) 273-1685

National Research
  & Chemical
14439 S. Avalon Blvd.
Gardena, CA 90248
(213) 515-1700

Safer Inc.
60 William St.
Wellesley, MA 02181
(617) 237-9660

Sierra Crop Protection
12101 Woodcrest
  Executive Way
St. Louis, MO 63141
(314) 275-7561

Southern Agricultural
  Insecticides Inc.
Box 429
Hendersonville, NC 28793
(704) 692-2233

Chemicals, antidesiccants

Easy Gardener Inc.
Box 21025
Waco, TX 78702
(800) 327-9462

Miller Chemical
  & Fertilizer Corp.
Box 333
Hanover, PA 17331
(717) 632-8921

PBI-Gordon Corp.
1217 W. 12th St.
Kansas City, MO 64101
(800) 821-7925

Rockland Chemical Co. Inc.
Box 809
West Caldwell, NJ 07006
(201) 575-1322

Sierra Crop Protection
12101 Woodcrest
  Executive Way
St. Louis, MO 63141
(314) 275-7561

Southern Agricultural
  Insecticides Inc.
Box 429
Hendersonville, NC 28793
(704) 692-2233

Wilt-Pruf Products Inc.
Box 4280
Greenwich, CT 06830
(203) 531-4740

Chemicals, antitranspirants

Aquatrols Corp. of America
1432 Union Ave.
Pennsauken, NJ 08110
(609) 665-1130

Bonide Chemical
2 Wurz Ave.
Yorkville, NY 13495
(315) 736-8231

Easy Gardener Inc.
Box 21025
Waco, TX 78702
(800) 327-9462

Lebanon Chemical Corp.
Box 180
Lebanon, PA 17042
(717) 273-1685

Miller Chemical
& Fertilizer Corp.
Box 333
Hanover PA 17331
(717) 632-8921

PBI-Gordon Corp.
1217 W. 12th St.
Kansas City, MO 64101
(800) 821-7925

Safer Inc.
60 William St.
Wellesley, MA 02181
(617) 237-9660

Sierra Crop Protection
12101 Woodcrest
Way
St. Louis, MO 63141
(314) 275-7561

Solar Sunstill
644 W. San Francisco
Santa Fe, NM 87501
(505) 982-8889

Southern Agricultural
Insecticides Inc.
Box 429
Hendersonville, NC 28793
(704) 692-2233

Wilt-Pruf Products Inc.
Box 4280
Greenwich, CT 06830
(203) 531-4740

Chemicals, bactericides

DelTek Inc.
Box 179
Pearland, TX 77588
(800) 367-3951

Merck & Co. Inc.
MSD Agvet Div.
Box 2000
Rahway, NJ 07065
(201) 855-4277

National Research
& Chemical
14439 S. Avalon Blvd.
Gardena, CA 90248
(213) 515-1700

OFE Intl. Inc.
12370 SW 130 St.
Miami, FL 33186
(305) 253-7080

Source Technology
Biologicals Inc.
3355 Hiawatha Ave. S.,
Ste. 122
Minneapolis, MN 55406
(612) 724-7102

Southern Agricultural
Insecticides Inc.
Box 429
Hendersonville, NC 28793
(704) 692-2233

Chemicals, botanicals

Bonide Chemical
2 Wurz Ave.
Yorkville, NY 13495
(315) 736-8231

Miller Chemical
& Fertilizer Corp.
Box 333
Hanover, PA 17331
(717) 632-8921

Chemicals, dyes, indicators

Lesco Inc.
20005 Lake Rd.
Rocky River OH 44116
(216) 333-9250

Lebanon Chemical Corp.
1600 E. Cumberland St. 180
Lebanon, PA 17042
(717) 273-1685

Southern Agricultural
Insecticides Inc.
Box 429
Hendersonville, NC 28793
(704) 692-2233

Chemicals, fumigants

Black Leaf Products Co.
1400 E. Touhy Ave.
Des Plaines, IL 60172
(708) 390-7070

Hendrix & Dail Inc.
Box 648
Greenville, NC 27835-0648
(919) 758-4263

Lebanon Chemical Corp.
Box 180
Lebanon, PA 17042
(717) 273-1685

Liquid Carbonic
135 S. La Salle St.
Chicago, IL 60603
(312) 855-2500

Plant Products Corp.
Box 1149
Vero Beach, FL 32960
(407) 567-7035

Southern Agricultural
Insecticides Inc.
Box 429
Hendersonville, NC 28793
(704) 692-2233

Wilbur-Ellis Co.
Box 47907
San Antonio, TX 78265-7907
(512) 227-5255

Chemicals, fungicides

Black Leaf Products Co.
1400 E. Touhy Ave.
Des Plaines, IL 60172
(708) 390-7070

Bonide Chemical
2 Wurz Ave.
Yorkville, NY 13495
(315) 736-8231

Ciba-Geigy Turf &
Ornamental Products
Box 18300
Greensboro, NC 27419
(919) 547-1160

DelTek Inc.
Box 179
Pearland, TX 77588
(800) 367-3951

DuPont Agricultural Products
Barley Mill Plaza
Wilmington, DE 19898
(302) 992-6173

Elanco Products Co.
Lily Corporate Ctr.
Indianapolis, IN 46285
(317) 276-2679

FMC Corp.
2000 Market St.
Philadelphia, PA 19103
(215) 299-6591

Fermenta Plant
Protection Co.
Box 8000
Mentor, OH 44061-8000
(216) 357-4100

Hendrix & Dail Inc.
Box 648
Greenville, NC 27835-0648
(919) 758-4263

Holtkamp Ghses. Inc.
Box 8158
Nashville, TN 37207
(615) 228-2683

Lesco Inc.
20005 Lake Rd.
Rocky River, OH 44116
(216) 333-9250

Lebanon Chemical Corp.
Box 180
Lebanon, PA 17042
(717) 273-1685

Merck & Co. Inc.
MSD Agvet Div.
Box 2000
Rahway, NJ 07065
(201) 855-4277

OFE Intl. Inc.
12370 SW 130 St.
Miami, FL 33186
(305) 253-7080

PBI-Gordon Corp.
1217 W. 12th St.
Kansas City, MO 64101
(800) 821-7925

Rhone-Poulenc Ag Co.
Box 12014
Research Triangle Park,
NC 27709
(919) 549-2000

Rockland Chemical Co. Inc.
Box 809
West Caldwell, NJ 07006
(201) 575-1322

Safer Inc.
60 William St.
Wellesley, MA 02181
(617) 237-9660

Sierra Crop Protection
12101 Woodcrest
Way
St. Louis, MO 63141
(314) 275-7561

Source Technology
Biologicals Inc.
3355 Hiawatha Ave. S.,
Ste. 122
Minneapolis, MN 55406
(612) 724-7102

Southern Agricultural
Insecticides Inc.
Box 429
Hendersonville, NC 28793
(704) 692-2233

Valent U.S.A. Corp.
Box 8025, Room 6051
Walnut Creek, CA 94596
(415) 256-2725

Wilbur-Ellis Co.
Box 47907
San Antonio, TX 78265-7907
(512) 227-5255

Chemicals, insecticides

Black Leaf Products Co.
1400 E. Touhy Ave.
Des Plaines, IL 60172
(708) 390-7070

Bonide Chemical
2 Wurz Ave.
Yorkville, NY 13495
(315) 736-8231

Ciba-Geigy Turf &
Ornamental Products
Box 18300
Greensboro, NC 27419
(919) 547-1160

DuPont Agricultural Products
Barley Mill Plaza
Wilmington, DE 19898
(302) 992-6173

FMC Corp.
2000 Market St.
Philadelphia, PA 19103
(215) 299-6591

Lesco Inc.
20005 Lake Rd.
Rocky River, OH 44116
(216) 333-9250

Lebanon Chemical Corp.
Box 180
Lebanon, PA 17042
(717) 273-1685

Merck & Co. Inc.
MSD Agvet Div.
Box 2000
Rahway, NJ 07065
(201) 855-4277

Midwest Spraying
& Supply Inc.
505 Brimhall Box 519
Long Lake, MN 55356
(612) 473-6499

Miller Chemical
& Fertilizer Corp.
Box 333
Hanover, PA 17331
(717) 632-8921

OFE Intl. Inc.
12370 SW 130 St.
Miami, FL 33186
(305) 253-7080

PBI-Gordon Corp.
1217 W. 12th St.
Kansas City, MO 64101
(800) 821-7925

Rhone-Poulenc Ag Co.
Box 12014
Research Triangle Park,
NC 27709
(919) 549-2000

Rockland Chemical Co. Inc.
Box 809
West Caldwell, NJ 07006
(201) 575-1322

Safer Inc.
60 William St.
Wellesley, MA 02181
(617) 237-9660

Sandoz Crop Protection Corp.
1300 E. Touhy Ave.
Des Plaines, IL 60018
(708) 699-1616

Schultz Co.
11730 Northline
St. Louis, MO 63043
(314) 567-4545

Sierra Crop Protection
12101 Woodcrest Way
St. Louis, MO 63141
(314) 275-7561

Southern Agricultural
Insecticides Inc.
Box 429
Hendersonville, NC 28793
(704) 692-2233

Trap-A-Fly
2701 Amelia Ave.
Panama City, FL 32405
(904) 769-0509

Valent U.S.A. Corp.
Box 8025, Room 6051
Walnut Creek, CA 94596
(415) 256-2725

Whitmire Research
35689 Tree Court
St. Louis, MO 63122
(314) 225-5371

Wilbur-Ellis Co.
Box 47907
San Antonio, TX 78265-7907
(512) 227-5255

Chemicals, miticides

Black Leaf Products Co.
1400 E. Touhy Ave.
Des Plaines, IL 60172
(708) 390-7070

Bonide Chemical
2 Wurz Ave.
Yorkville, NY 13495
(315) 736-8231

DuPont Agricultural Products
Barley Mill Plaza
Wilmington, DE 19898
(302) 992-6173

FMC Corp.
2000 Market St.
Philadelphia, PA 19103
(215) 299-6591

Lesco Inc.
20005 Lake Rd.
Rocky River, OH 44116
(216) 333-9250

Lebanon Chemical Corp.
Box 180
Lebanon, PA 17042
(717) 273-1685

Merck & Co. Inc.
MSD Agvet Div.
Box 2000
Rahway, NJ 07065
(201) 855-4277

PBI-Gordon Corp.
1217 W. 12th St.
Kansas City, MO 64101
(800) 821-7925

Rhone-Poulenc Ag Co.
Box 12014
Research Triangle Park,
  NC 27709
(919) 549-2000

Rockland Chemical Co. Inc.
Box 809
West Caldwell, NJ 07006
(201) 575-1322

Safer Inc.
60 William St.
Wellesley, MA 02181
(617) 237-9660

Sandoz Crop Protection Corp.
1300 E. Touhy Ave.
Des Plaines, IL 60018
(708) 699-1616

Southern Agricultural
  Insecticides Inc.
Box 429
Hendersonville, NC 28793
(704) 692-2233

Uniroyal Chemical
World HQ-AG Group
Middlebury, CT 06749
(203) 573-2000

Valent U.S.A. Corp.
Box 8025, Room 6051
Walnut Creek, CA 94596
(415) 256-2725

Whitmire Research
35689 Tree Court
St. Louis, MO 63122
(314) 225-5371

Wilbur-Ellis Co.
Box 47907
San Antonio, TX 78265-7907
(512) 227-5255

**Chemicals, snail and
slug control**

Black Leaf Products Co.
1400 E. Touhy Ave.
Des Plaines, IL 60172
(708) 390-7070

Bonide Chemical
2 Wurz Ave.
Yorkville, NY 13495
(315) 736-8231

Lebanon Chemical Corp.
Box 180
Lebanon, PA 17042
(717) 273-1685

OFE Intl. Inc.
12370 SW 130 St.
Miami, FL 33186
(305) 253-7080

PBI-Gordon Corp.
1217 W. 12th St.
Kansas City, MO 64101
(800) 821-7925

Seabright Ltd.
4026 Harlan St.
Emeryville, CA 94608
(415) 655-3126

Southern Agricultural
  Insecticides Inc.
Box 429
Hendersonville, NC 28793
(704) 692-2233

Valent U.S.A. Corp.
Box 8025, Room 6051
Walnut Creek, CA 94596
(415) 256-2725

Vigoro Industries Inc.
2007 W. Hwy. 50
Box 4139
Fairview Heights, IL 62208-
  2928
(618) 624-5522

Wilbur-Ellis Co.
Box 47907
San Antonio, TX 78265-7907
(512) 227-5255

**Chemicals,
spreader-stickers**

Black Leaf Products Co.
1400 E. Touhy Ave.
Des Plaines, IL 60172
(708) 390-7070

Bonide Chemical
2 Wurz Ave.
Yorkville, NY 13495
(315) 736-8231

Lebanon Chemical Corp.
Box 180
Lebanon, PA 17042
(717) 273-1685

Lesco Inc.
20005 Lake Rd.
Rocky River, OH 44116
(216) 333-9250

Miller Chemical
  & Fertilizer Corp.
Box 333
Hanover PA 17331
(717) 632-8921

National Research
  & Chemical
14439 S. Avalon Blvd.
Gardena, CA 90248
(213) 515-1700

OFE Intl. Inc.
12370 SW 130 St.
Miami, FL 33186
(305) 253-7080

PBI-Gordon Corp.
1217 W. 12th St.
Kansas City, MO 64101
(800) 821-7925

Rockland Chemical Co. Inc.
Box 809
West Caldwell, NJ 07006
(201) 575-1322

Safer Inc.
60 William St.
Wellesley, MA 02181
(617) 237-9660

Sierra Crop Protection
12101 Woodcrest
  Way
St. Louis, MO 63141
(314) 275-7561

Southern Agricultural
  Insecticides Inc.
Box 429
Hendersonville, NC 28793
(704) 692-2233

Valent U.S.A. Corp.
Box 8025, Room 6051
Walnut Creek, CA 94596
(415) 256-2725

Wilbur-Ellis Co.
Box 47907
San Antonio, TX 78265-7907
(512) 227-5255

Chemicals, wetting agents

Aglukon
50 N. Harrison Ave.
Congers, NY 01920
(914) 268-2122

Aquatrols Corp. of America
1432 Union Ave.
Pennsauken, NJ 08110
(609) 665-1130

Ball Seed Co.
622 Town Rd.
West Chicago, IL 60185
(708) 231-3500

Broadleaf Industries
7014 Manya Circle
San Diego, CA 92154
(619) 424-7880

EnP Inc.
Box 218
Mendota, IL 61342
(815) 539-7471

Industrial Svcs. Intl. Inc.
Box 10834
Bradenton, FL 34282-0834
(813) 792-7778

Lebanon Chemical Corp.
1600 E. Cumberland St.
Box 180
Lebanon, PA 17042
(717) 273-1685

Lesco Inc.
20005 Lake Rd.
Rocky River, OH 44116
(216) 333-9250

Miller Chemical
& Fertilizer Corp.
Box 333
Hanover PA 17331
(717) 632-8921

Moisture Mizer Div.
Multiple Concepts Inc.
Box 4248
Chattanooga, TN 37405
(615) 266-3967

PBI-Gordon Corp.
1217 W. 12th St.
Kansas City, MO 64101
(800) 821-7925

Sierra Crop Protection
12101 Woodcrest
Way
St. Louis, MO 63141
(314) 275-7561

Southern Agricultural
Insecticides Inc.
Box 429
Hendersonville, NC 28793
(704) 692-2233

Valent U.S.A. Corp.
Box 8025, Room 6051
Walnut Creek, CA 94596
(415) 256-2725

Wilbur-Ellis Co.
Box 47907
San Antonio, TX 78265-7907
(512) 227-5255

Computers, consultants

AgTech Consultants Inc.
1280 Terminal Way, Ste. 24
Reno, NV 89502
(702) 322-5900

Agri-Land Data Systems Ltd.
160 King St.
Thorndale, ON NOM 2PO
CANADA
(519) 461-0843

Argus Control Systems Ltd.
10-1480 Foster St.
White Rock, BC V4B 3X7
CANADA
(604) 536-3171

Condor Computing Inc.
Box 17276
Huntsville, AL 35810
(205) 852-4490

Datarose Inc.
Box 129
Bowling Green, KY 42101
(502) 781-3282

Datasphere Computer
Systems Inc.
6443 SW Beaverton-
Hillsdale Hwy., Ste. 305
Portland, OR 97221
(503) 297-9035

New Era/Cressmark
Business Automation
Drawer X
Smithtown, NY 11787
(516) 234-3917

Q-Com Inc.
2050 S. Grand Ave.
Santa Ana, CA 92705
(714) 540-6123

Computers, hardware

ADS Software Inc.
707 5th St. NE
Roanoke, VA 24016
(703) 344-6818

AgTech Consultants Inc.
1280 Terminal Way, Ste. 24
Reno, NV 89502
(702) 322-5900

Agri-Land Data
Systems Ltd.
160 King St.
Thorndale, ON N0M 2PO
CANADA
(519) 461-0843

Condor Computing Inc.
Box 17276
Huntsville, AL 35810
(205) 852-4490

DACE Inc.
1937 High St.
Longwood, FL 32750
(407) 321-7771

Datarose Inc.
Box 129
Bowling Green, KY 42101
(502) 781-3282

Datasphere Computer
Systems Inc.
6443 SW Beaverton-
Hillsdale Hwy., Ste. 305
Portland, OR 97221
(503) 297-9035

Economy Label Sales Co. Inc.
Box 350
Daytona Beach, FL 32015
(904) 253-4741

Q-Com Inc.
2050 S. Grand Ave.
Santa Ana, CA 92705
(714) 540-6123

Thornton Computer Mgt.
Systems
424 E. US 22
Maineville, OH 45039
(513) 683-8100

Troy Hygro-Systems Inc.
4096 Hwy. ES
East Troy, WI 53120
(414) 642-5928

VitroTech Corp.
701 Devonshire Dr., #C-24
Champaign, IL 61820
(217) 352-7190

Wadsworth Control Systems
5541 Marshall St.
Arvada, CO 80002
(303) 424-4461

**Computers, software,
business management**

ADS Software Inc.
707 5th St. NE
Roanoke, VA 24016
(703) 344-6818

AgTech Consultants Inc.
1280 Terminal Way, Ste. 24
Reno, NV 89502
(702) 322-5900

Agri-Land Data Systems Ltd.
160 King St.
Thorndale, ON N0M 2PO
CANADA
(519) 461-0843

Condor Computing Inc.
Box 17276
Huntsville, AL 35810
(205) 852-4490

Datarose Inc.
Box 129
Bowling Green, KY 42101
(502) 781-3282

Datasphere Computer
Systems Inc.
6443 SW Beaverton-
Hillsdale Hwy, Ste. 305
Portland, OR 97221
(503) 297-9035

Great Plains Software Inc.
1701 38th St. SW
Fargo, ND 58103
(701) 281-0550

Micro Vane Inc.
8135 Cox's Dr.
Kalamazoo, MI 49002
(616) 329-0188

New Era/Cressmark
Business Automation
Drawer X
Smithtown, NY 11787
(516) 234-3917

Priva Computers Inc.
Box 110
Vineland Station, ON L0R
2E0
CANADA
(416) 562-7351

Thornton Computer Mgt.
Systems
424 E. US 22
Maineville, OH 45039
(513) 683-8100

**Computers, software,
environmental control**

Aerovent Fan & Eqt. Inc.
929 Terminal Rd.
Lansing, MI 48906
(517) 323-2930

Argus Control Systems Ltd.
10-1480 Foster St.
White Rock, BC V4B 3X7
CANADA
(604) 536-3171

Automata Inc.
19393 Redberry Rd.
Grass Valley, CA 95945
(916) 273-0380

Bio-Energy Systems Inc.
Box 191
Ellenville, NY 12428
(914) 647-6700

DACE Inc.
1937 High St.
Longwood, FL 32750
(407) 321-7771

Economy Label Sales Co. Inc.
Box 350
Daytona Beach, FL 32015
(904) 253-4741

Egor-Hoppmann Corp.
Box 601
Chantilly, VA 22021
(703) 631-2700

Gerhart Inc.
6346 Avon Belden Rd.
North Ridgeville OH 44039
(216) 327-8056

Lander Control Systems Inc.
RR 1
Orangeville, ON L9W 2Y8
CANADA
(519) 941-9880

Midwest Trading,
GROmaster Div.
Box 384
St. Charles, IL 60174
(708) 742-1840

Neogen Food Tech
620 Lesher Pl.
Lansing, MI 48912
(517) 372-9200

Oglevee Computer Div.
150 Oglevee Rd.
Connellsville, PA 15425
(412) 628-8360

Pacific Controlled
Environments
Box 26
Lake Elsinore, CA 92330
(714) 674-1556

Precision Growth Systems
3350 Scott Blvd., Bldg. 61
Santa Clara, CA 95054
(408) 727-6256

Priva Computers Inc.
Box 110
Vineland Station ON L0R
2E0
CANADA
(416) 562-7351

Q-Com Inc.
2050 S. Grand Ave.
Santa Ana, CA 92705
(714) 540-6123

Remote Measurement
Systems
2633 Eastlake Ave. E.
Ste. 200
Seattle, WA 98102
(206) 328-2255

Specialty Products
& Svcs. Corp.
Box 20909
San Jose, CA 95160
(408) 997-6100

Troy Hygro-Systems Inc.
4096 Hwy. ES
East Troy, WI 53120
(414) 642-5928

V & V Noordland Inc.
Box 739
Medford, NY 11763
(516) 698-2300

VitroTech Corp.
701 Devonshire Dr. #C-24
Champaign, IL 61820
(217) 352-7190

Wadsworth Control Systems
5541 Marshall St.
Arvada, CO 80002
(303) 424-4461

Condensate control

Agra Tech Inc.
2131 Piedmont Way
Pittsburg, CA 94565
(415) 432-3399

Pacific Controlled
Environments
Box 26
Lake Elsinore, CA 92330
(714) 674-1556

Specialty Products
& Svcs. Corp.
Box 20909
San Jose, CA 95160
(408) 997-6100

**Consultants,
environmental control**

Aerovent Fan & Eqt. Inc.
929 Terminal Rd.
Lansing, MI 48906
(517) 323-2930

Agro Dynamics
Building 3
Navy Yard, Brooklyn, NY
11205
(718) 596-3042

Argus Control Systems Ltd.
10-1480 Foster St.
White Rock, BC V4B 3X7
CANADA
(604) 536-3171

Biotherm Engineering
Box 6007
Petaluma, CA 94953
(707) 763-4444

Lander Control Systems Inc.
RR 1, Orangeville
ON L9W 2Y8
CANADA
(519) 941-9880

Albert J. Lauer Inc.
16700 Chippendale Ave. W
(Hwy. 3)
Rosemount, MN 55068
(612) 423-1651

Marbil Engineering Assoc.
6910 Furman Pkwy.
Riverdale, MD 20737
(301) 459-1607

Pacific Controlled
Environments
Box 26
Lake Elsinore, CA 92330
(714) 674-1556

Phonetics Inc.
101 State Rd.
Media, PA 19063
(215) 565-8520

Precision Growth Systems
3350 Scott Blvd, Bldg. 61
Santa Clara, CA 95054
(408) 727-6256

Dehumidifiers

Ball Seed Co.
622 Town Rd.
West Chicago, IL 60185
(708) 231-3500

Dansco Distributors
4442 27th Ave. W.
Seattle, WA 98199
(206) 282-7282

Seedburo Eqt. Co.
1022 W. Jackson Blvd.
Chicago, IL 60607
(312) 738-3700

**Electrical conductivity
(EC) meters**

Agro Dynamics
Building 3, Navy Yard
Brooklyn, NY 11205
(718) 596-3042

H.E. Anderson Co. Inc.
Box 1006
Muskogee, OK 74402-1006
(918) 687-4426

Engineered Systems &
Designs
3 S. Tatnall St.
Wilmington, DE 19801
(302) 571-1195

J-M Trading Corp. Inc.
241 Frontage Rd., Ste. 47
Burr Ridge, IL 60521
(708) 655-3305

Kel Instruments Co. Inc.
Box 2174
Vineyard Haven, MA 02568
(508) 693-7798

LaMotte Chemical Products
Co.
128 N. Queen St.
Chestertown, MD 21620
(301) 778-3100

Ted Mahr Supply Inc.
Box 150310
Cape Coral, FL 33915
(813) 574-2214

Metex Corp. Ltd.
12 Penn Dr., Unit 1
Weston, ON M9L 2A9
CANADA
(416) 749-1210

Midwest Trading,
GROmaster Div.
Box 384
St. Charles, IL 60174
(708) 742-1840

Myron L Co.
6231 Yarrow Dr., Ste. C
Carlsbad, CA 92009
(619) 438-2021

Troy Hygro-Systems Inc.
4096 Hwy. ES
East Troy, WI 53120
(414) 642-5928

Environmental controllers

AAA Associates Inc.
1445 S. 3rd St.
Niles, MI 49120
(616) 684-4073

Acme Engineering
& Mfg. Corp.
Box 978
Muskogee, OK 74402
(918) 682-7791

Aerovent Fan & Eqt. Inc.
929 Terminal Rd.
Lansing, MI 48906
(517) 323-2930

Argus Control Systems Ltd.
10-1480 Foster St.
White Rock, BC V4B 3X7
CANADA
(604) 536-3171

Atomizing Systems Inc.
1 Hollywood Ave.
Ho-Ho-Kus, NJ 07423
(201) 447-1222

Automata Inc.
19393 Redberry Rd.
Grass Valley, CA 95945
(916) 273-0380

Ball Seed Co.
622 Town Rd.
West Chicago, IL 60185
(708) 231-3500

Biotherm Engineering
Box 6007
Petaluma, CA 94953
(707) 763-4444

Clover Ghses. by Elliott Inc.
Box 789
Smyrna, TN 37167
(615) 459-3863

Egor-Hoppmann Corp.
Box 601
Chantilly, VA 22021
(703) 631-2700

Gerhart Inc.
6346 Avon Belden Rd.
North Ridgeville, OH 44039
(216) 327-8056

Lander Control Systems Inc.
RR 1
Orangeville, ON L9W 2Y8
CANADA
(519) 941-9880

Midwest Trading,
  GROmaster Div.
Box 384
St. Charles, IL 60174
(708) 742-1840

Neogen Food Tech
620 Lesher Pl.
Lansing, MI 48912
(517) 372-9200

Pacific Controlled
  Environments
Box 26
Lake Elsinore, CA 92330
(714) 674-1556

Phonetics Inc.
101 State Rd.
Media, PA 19063
(215) 565-8520

Precision Growth Systems
3350 Scott Blvd, Bldg. 61
Santa Clara, CA 95054
(408) 727-6256

Q-Com Inc.
2050 S. Grand Ave.
Santa Ana, CA 92705
(714) 540-6123

Remote Measurement
  Systems
2633 Eastlake Ave. E.
  Ste. 200
Seattle, WA 98102
(206) 328-2255

Reznor
McKinley Ave.
Mercer, PA 16137
(412) 662-4400

Sera Solar Corp.
3151 Jay St.
Santa Clara, CA 95054
(408) 727-3747

United Ghse. Systems Inc.
708 Washington St.
Edgerton, WI 53534
(800) 433-6834

Vary Industries
Box 160
Grimsby, ON L3M 4N6
CANADA
(416) 945-9691

VitroTech Corp.
701 Devonshire Dr., #C-24
Champaign, IL 61820
(217) 352-7190

Wadsworth Control Systems
5541 Marshall St.
Arvada, CO 80002
(303) 424-4461

Winandy Ghse. Co. Inc.
2211 Peacock Rd.
Richmond, IN 47374
(317) 935-2111

Fans, fan-jet system

AAA Associates Inc.
1445 S. 3rd St.
Niles, MI 49120
(616) 684-4073

Acme Engineering
  & Mfg. Corp.
Box 978
Muskogee, OK 74402
(918) 682-7791

American Coolair Corp.
Box 2300
Jacksonville, FL 32203
(904) 389-3646

Ball Seed Co.
622 Town Rd.
West Chicago, IL 60185
(708) 231-3500

Barbrook Inc.
Box 6601
Shreveport, LA 71136-6601
(318) 868-3626

Clover Ghses. by Elliott Inc.
Box 789
Smyrna, TN 37167
(615) 459-3863

Conley's Mfg. & Sales
4344 E. Mission Blvd.
Pomona, CA 91766
(714) 627-0981

De Cloet Ghses.,
  Div. Daylight Mfg. Ltd.
RR 1
Simcoe ON N3Y 4J9
CANADA
(519) 582-3081

Growers Ghse. Supplies
Box 83
Vineland Station, ON L0R
  2E0
CANADA
(416) 562-7341

Harnois Industries Inc.
1044 Principale
St. Thomas (Joliette), PQ
  J0K 3L0
CANADA
(514) 756-1041

Ken-Bar Inc.
24 Gould St.
Reading, MA 01867
(617) 944-0003

Albert J. Lauer Inc.
16700 Chippendale Ave. W
  (Hwy. 3)
Rosemount, MN 55068
(612) 423-1651

National Ghse. Co.
Box 100
Pana, IL 62557
(217) 562-9333

Penn Ventilator Co. Inc.
Red Lion & Gantry Roads
Philadelphia, PA 19025
(215) 464-8900

Poly Grower Ghse. Co.
Box 359
Muncy, PA 17756
(717) 546-3216

Roundhouse by Groce
Box 1744
Cleveland, TX 77327
(713) 592-7474

X.S. Smith Inc.
Drawer X
Red Bank, NJ 07701
(201) 222-4600

Troy Hygro-Systems Inc.
4096 Hwy. ES
East Troy, WI 53120
(414) 642-5928

United Ghse. Systems Inc.
708 Washington St.
Edgerton, WI 53534
(800) 433-6834

Vary Industries
Box 160
Grimsby, ON L3M 4N6
CANADA
(416) 945-9691

Westbrook Ghse. Systems
Ltd.
270 Hunter Rd.
Box 99
Grimsby, ON L3M 4G1
CANADA
(416) 945-4111

Winandy Ghse. Co. Inc.
2211 Peacock Rd.
Richmond, IN 47374
(317) 935-2111

**Fans, greenhouse cooling
and ventilating**

AAA Associates Inc.
1445 S. 3rd St.
Niles, MI 49120
(616) 684-4073

Aalbrecht, Luc,
Tuinbouwtechniek B.V.
Franklinstraat 4/Postbus 71,
2690 AB 's
Gravenzande, HOLLAND
01748-17101
FAX 01748-12901

Acme Engineering
& Mfg. Corp.
Box 978
Muskogee, OK 74402
(918) 682-7791

Aerovent Fan & Eqt. Inc.
929 Terminal Rd.
Lansing, MI 48906
(517) 323-2930

Agri Inc.
Box 577
Broadway, NC 27505
(919) 258-9113

American Coolair Corp.
Box 2300
Jacksonville, FL 32203
(904) 389-3646

Ball Seed Co.
622 Town Rd.
West Chicago, IL 60185
(708) 231-3500

B&W Ghse. Construction
Ltd.
Box 307
Aldergrove, BC V0X 1A0
CANADA
(604) 852-5848

Barbrook Inc.
6135 Linwood Ave.
Box 6601
Shreveport, LA 71136-6601
(318) 868-3626

Baumac Intl.
1500 Crafton Ave.
Mentone, CA 92359
(714) 794-7631

Clover Ghses. by Elliott Inc.
200 Weakley Lane
Box 789
Smyrna, TN 37167
(615) 459-3863

Conley's Mfg. & Sales
4344 E. Mission Blvd.
Pomona, CA 91766
(714) 627-0981

DACE Inc.
1937 High St.
Longwood, FL 32750
(407) 321-7771

Dansco Distributors
4442 27th Ave. W.
Seattle, WA 98199
(206) 282-7282

De Cloet Ghses.
Div. Daylight Mfg. Ltd.
RR 1
Simcoe, ON N3Y 4J9
CANADA
(519) 582-3081

Godro Inc.
Box 1682
Bloomington, IL 61701-1682
(309) 829-4353

Growers Ghse. Supplies
N. Service Rd.
Box 83
Vineland Station, ON L0R
2E0
CANADA
(416) 562-7341

Harnois Industries Inc.
1044 Principale
St. Thomas (Joliette), PQ
J0K 3L0
CANADA
(514) 756-1041

Jaderloon Co. Inc.
Box 685
Irmo, SC 29063
(803) 798-4000

Jaybird Mfg. Inc.
Rt. 322, RD 1
Box 489A
Centre Hall, PA 16828
(814) 364-1810

Albert J. Lauer Inc.
16700 Chippendale Ave. W
(Hwy. 3)
Rosemount, MN 55068
(612) 423-1651

Ludy Ghse. Mfg. Corp.
Box 141
New Madison, OH 45346
(513) 996-1921

Micro Cool
653 Commercial Rd., Ste. 1
Palm Springs, CA 92262
(619) 322-1111

National Ghse. Co.
Box 100
Pana, IL 62557
(217) 562-9333

Penn Ventilator Co. Inc.
Red Lion & Gantry Roads
Philadelphia, PA 19025
(215) 464-8900

Poly Grower Ghse. Co.
Box 359
Muncy, PA 17756
(717) 546-3216

Roundhouse by Groce
Box 1744
Cleveland, TX 77327
(713) 592-7474

X.S. Smith Inc.
Drawer X
Red Bank, NJ 07701
(201) 222-4600

Specialty Products
& Svcs. Corp.
Box 20909
San Jose, CA 95160
(408) 997-6100

Stuppy Ghse. Mfg. Inc.
Box 12456
North Kansas City, MO 64116
(816) 842-3071

System USA Inc.
Box 777
Watsonville, CA 95076
(408) 722-1188

Troy Hygro-Systems Inc.
4096 Hwy. ES
East Troy, WI 53120
(414) 642-5928

Tuscarora Electric Mfg.
Co. Inc.
Hilltop Dr., RD 3
Tunkhannock, PA 18657
(717) 836-2101

United Ghse. Systems Inc.
708 Washington St.
Edgerton, WI 53534
(800) 433-6834

V & V Noordland Inc.
Box 739
Medford, NY 11763
(516) 698-2300

Vary Industries
Box 160
Grimsby, ON L3M 4N6
CANADA
(416) 945-9691

Westbrook Ghse. Systems
  Ltd.
Box 99
Grimsby, ON L3M 4G1
CANADA
(416) 945-4111

Winandy Ghse. Co. Inc.
2211 Peacock Rd.
Richmond, IN 47374
(317) 935-2111

Fans, horizontal air flow

AAA Associates Inc.
1445 S. 3rd St.
Niles, MI 49120
(616) 684-4073

Aalbrecht, Luc,
  Tuinbouwtechniek B.V.
Franklinstraat 4/Postbus 71,
  2690 AB 's
Gravenzande, HOLLAND
01748-17101
FAX 01748-12901

Acme Engineering
  & Mfg. Corp.
Box 978
Muskogee, OK 74402
(918) 682-7791

Ball Seed Co.
622 Town Rd.
West Chicago, IL 60185
(708) 231-3500

Barbrook Inc.
Box 6601
Shreveport, LA 71136-6601
(318) 868-3626

DACE Inc.
1937 High St.
Longwood, FL 32750
(407) 321-7771

Dansco Distributors
4442 27th Ave. W.
Seattle, WA 98199
(206) 282-7282

De Cloet Ghses.,
  Div. Daylight Mfg. Ltd.
RR 1
Simcoe, ON N3Y 4J9
CANADA
(519) 582-3081

Godro Inc.
Box 1682
Bloomington, IL 61701-1682
(309) 829-4353

Growers Ghse. Supplies
Box 83
Vineland Station, ON L0R
  2E0
CANADA
(416) 562-7341

Jaderloon Co. Inc.
Box 685
Irmo, SC 29063
(803) 798-4000

Johnson Gas Appliance Co.
520 E. Ave., NW
Cedar Rapids, IA 52405
(319) 365-5267

National Ghse. Co.
Box 100
Pana, IL 62557
(217) 562-9333

Poly Grower Ghse. Co.
Box 359
Muncy, PA 17756
(717) 546-3216

Priva Computers Inc.
Box 110
Vineland Station, ON L0R
  2E0
CANADA
(416) 562-7351

Troy Hygro-Systems Inc.
4096 Hwy. ES
East Troy, WI 53120
(414) 642-5928

V & V Noordland Inc.
Box 739
Medford, NY 11763
(516) 698-2300

Vary Industries
Box 160
Grimsby, ON L3M 4N6
CANADA
(416) 945-9691

Westbrook Ghse. Systems
  Ltd.
Box 99
Grimsby, ON L3M 4G1
CANADA
(416) 945-4111

Winandy Ghse. Co. Inc.
2211 Peacock Rd.
Richmond, IN 47374
(317) 935-2111

Fog systems

AAA Associates Inc.
1445 S. 3rd St.
Niles, MI 49120
(616) 684-4073

Atomizing Systems Inc.
1 Hollywood Ave.
Ho-Ho-Kus, NJ 07423
(201) 447-1222

Ball Seed Co.
622 Town Rd.
West Chicago, IL 60185
(708) 231-3500

Baumac Intl.
1500 Crafton Ave.
Mentone, CA 92359
(714) 794-7631

Clover Ghses. by Elliott Inc.
Box 789
Smyrna, TN 37167
(615) 459-3863

DACE Inc.
1937 High St.
Longwood, FL 32750
(407) 321-7771

Jaybird Mfg. Inc.
Box 489A
Centre Hall, PA 16828
(814) 364-1810

Mee Industries
4443 N. Rowland Ave.
El Monte, CA 91731
(818) 350-4180

Micro Cool
653 Commercial Rd., Ste. 1
Palm Springs, CA 92262
(619) 322-1111

Midwest Spraying
  & Supply Inc.
505 Brimhall
Box 519
Long Lake, MN 55356
(612) 473-6499

Midwest Trading,
  GROmaster Div.
Box 384
St. Charles, IL 60174
(708) 742-1840

Multi-Tech Co. Ltd.
P.O. Box 2119
Taichung City, Taiwan,
R.O.C. 886-4-3260001

National Ghse. Co.
Box 100
Pana, IL 62557
(217) 562-9333

Plant Products Corp.
Box 1149
Vero Beach, FL 32960
(407) 567-7035

R L Corp.
1000 Foreman Rd.
Lowell, MI 49331
(616) 897-9211

Spraying Systems Co.
N. Ave. at Schmale Road
Wheaton, IL 60188
(708) 665-5000

Vary Industries
Box 160
Grimsby, ON L3M 4N6
CANADA
(416) 945-9691

Westbrook Ghse. Systems
Ltd.
Box 99
Grimsby, ON L3M 4G1
CANADA
(416) 945-4111

W.A. Westgate Co. Inc.
Box 445
Davis, CA 95617
(916) 753-2954

**Insect traps,
yellow sticky cards**

Automata Inc.
19393 Redberry Rd.
Grass Valley, CA 95945
(916) 273-0380

DeWill Inc.
766-68 Industrial Dr.
Elmhurst, IL 60126
(708) 941-7210

Hort Services Inc.
Box 327
Geneva IL 60134
(708) 232-8184

Phero Tech Inc.
1140 Clark Dr.
Vancouver, BC V5L 3K3
CANADA
(604) 255-7381

Safer Inc.
60 William St.
Wellesley, MA 02181
(617) 237-9660

Seabright Ltd.
4026 Harlan St.
Emeryville, CA 94608
(415) 655-3126

Trap-A-Fly
2701 Amelia Ave.
Panama City, FL 32405
(904) 769-0509

**Laboratory testing services,
water**

Agro Dynamics
Building 3, Navy Yard
Brooklyn, NY 11205
(718) 596-3042

Agro Svcs. Intl. Inc.
215 E. Michigan Ave.
Orange City, FL 32763
(904) 775-6601

Conrad Fafard Inc.
Box 3190
Springfield, MA 01101
(413) 786-4343

Dansco Distributors
4442 27th Ave. W.
Seattle, WA 98199
(206) 282-7282

Grace-Sierra
6656 Grant Way
Allentown, PA 18106
(215) 395-7104

National Research
& Chemical
14439 S. Avalon Blvd.
Gardena, CA 90248
(213) 515-1700

Thornton Laboratories Inc.
1145 E. Cass St.
Tampa, FL 33602
(813) 223-9702

Light meters

Dorothy Biddle Service
Dept. G
Greeley, PA 18425-9799
(717) 226-3239

Dansco Distributors
4442 27th Ave. W.
Seattle, WA 98199
(206) 282-7282

Midwest Trading,
GROmaster Div.
Box 384
St. Charles, IL 60174
(708) 742-1840

**Pesticide applicators,
dusters**

Atomizing Systems Inc.
1 Hollywood Ave.
Ho-Ho-Kus, NJ 07423
(201) 447-1222

H. D. Hudson Mfg. Co.
500 N. Michigan Ave.
Chicago, IL 60611
(312) 644-2830

Midwest Spraying
& Supply Inc.
505 Brimhall Box 519
Long Lake, MN 55356
(612) 473-6499

Plant Products Corp.
Box 1149
Vero Beach FL 32960
(407) 567-7035

R L Corp.
1000 Foreman Rd.
Lowell, MI 49331
(616) 897-9211

M.K. Rittenhouse & Sons Ltd.
4th Ave. RR 3
St. Catharines, ON L2R 6P9
CANADA
(416) 684-8122

Soil-Tec Inc.
Box 59413
Dallas, TX 75229
(214) 263-0142

**Pesticide applicators,
foamers**

Midwest Spraying
& Supply Inc.
505 Brimhall Box 519
Long Lake, MN 55356
(612) 473-6499

**Pesticide applicators,
foggers**

Atomizing Systems Inc.
1 Hollywood Ave.
Ho-Ho-Kus, NJ 07423
(201) 447-1222

Ball Seed Co.
622 Town Rd.
West Chicago, IL 60185
(708) 231-3500

The Dramm Co.
Box 528
Manitowoc, WI 54220
(414) 684-0227

Midwest Spraying
& Supply Inc.
505 Brimhall Box 519
Long Lake, MN 55356
(612) 473-6499

R L Corp.
1000 Foreman Rd.
Lowell, MI 49331
(616) 897-9211

Soil-Tec Inc.
Box 59413
Dallas, TX 75229
(214) 263-0142

**Pesticide applicators,
soil fumigants**

Hendrix & Dail Inc.
803 Industrial Blvd Box 648
Greenville, NC 27835-0648
(919) 758-4263

**Pesticide applicators,
sprayers**

Agrotec Inc.
Box 49
Pendleton, NC 27862
(919) 585-1222

Atomizing Systems Inc.
1 Hollywood Ave.
Ho-Ho-Kus, NJ 07423
(201) 447-1222

Ball Seed Co.
622 Town Rd.
West Chicago, IL 60185
(708) 231-3500

Davis & Associates Inc.
802 Lingco Dr.
Richardson, TX 75081
(214) 234-5422

The Dramm Co.
Box 528
Manitowoc, WI 54220
(414) 684-0227

D.I.G.
16216 Raymer St.
Van Nuys, CA 91406
(818) 989-5999

Electrostatic Spraying
Systems Inc.
209 Gary Ave.
Selma, AL 36701
(205) 872-7201

H.D. Hudson Mfg. Co.
500 N. Michigan Ave.
Chicago, IL 60611
(312) 644-2830

Lesco Inc.
20005 Lake Rd.
Rocky River, OH 44116
(216) 333-9250

Mantis Mfg. Co.
1458 County Line Rd.
Huntingdon Valley, PA 19006
(215) 355-9700

Master Mfg. Co. Inc.
Box 3806
Sioux City, IA 51102
(712) 258-0108

Midwest Spraying
& Supply Inc.
505 Brimhall Box 519
Long Lake, MN 55356
(612) 473-6499

OFE Intl. Inc.
12370 SW 130 St.
Miami, FL 33186
(305) 253-7080

PBI-Gordon Corp.
1217 W. 12th St.
Kansas City, MO 64101
(800) 821-7925

R L Corp.
1000 Foreman Rd.
Lowell, MI 49331
(616) 897-9211

M.K. Rittenhouse & Sons Ltd.
4th Ave. RR3
St. Catharines, ON L2R 6P9
CANADA
(416) 684-8122

Siebring Mfg. Co.
303 S. Main
George, IA 51237
(712) 475-3317

Soil-Tec Inc.
Box 59413
Dallas, TX 75229
(214) 263-0142

True Friends Garden Tools
Inc.
Box 1278
Cumming, GA 30130
(404) 887-7815

Wilbur-Ellis Co.
Box 47907
San Antonio, TX 78265-7907
(512) 227-5255

Charles Wolfe & Assoc. Inc.
Box 282
Cumming, GA 30130
(404) 887-7815

**Pesticide applicators,
spreaders**

Agri-Fab Inc.
Box 903
Sullivan, IL 61951
(217) 728-8388

Gandy Co.
528 Gandrud Rd.
Owatonna, MN 55060
(507) 451-5430

Lesco Inc.
20005 Lake Rd.
Rocky River, OH 44116
(216) 333-9250

Midwest Spraying
& Supply Inc.
505 Brimhall Box 519
Long Lake, MN 55356
(612) 473-6499

PBI-Gordon Corp.
1217 W. 12th St.
Kansas City, MO 64101
(800) 821-7925

Spyker Spreaders
810 W. Main St.
North Manchester, IN 46962
(219) 982-8105

Pesticide safety equipment

Airstream Dust
& Chemical Helmets
Box 975
Elbow Lake, MN 56531
(218) 685-4457

Argus Supply Co.
Box 689
Roseville, MI 48066
(313) 774-8900

DuPont Agricultural Products
Barley Mill Plaza
Wilmington, DE 19898
(302) 992-6173

Gempler's
Box 270-75B
Mount Horeb, WI 53572
(608) 437-4883

Hendrix & Dail Inc.
Box 648
Greenville, NC 27835-0648
(919) 758-4263

Lesco Inc.
20005 Lake Rd.
Rocky River, OH 44116
(216) 333-9250

Midwest Spraying
& Supply Inc.
505 Brimhall Box 519
Long Lake, MN 55356
(612) 473-6499

Southern Agricultural
Insecticides Inc.
Box 429
Hendersonville, NC 28793
(704) 692-2233

Specialty Products
& Svcs. Corp.
Box 20909
San Jose, CA 95160
(408) 997-6100

The St. George Co. Ltd.
St. George, ON N0E 1N0
CANADA
(519) 442-2046

pH meters

Agro Dynamics
Building 3 Navy Yard
Brooklyn, NY 11205
(718) 596-3042

Dorothy Biddle Service
Dept. G
Greeley, PA 18425-9799
(717) 226-3239

Blackmore Transplanter Co.
10800 Blackmore Ave.
Belleville, MI 48111
(313) 483-8661

Dansco Distributors
4442 27th Ave. W.
Seattle, WA 98199
(206) 282-7282

Engineered Systems &
Designs
3 S. Tatnall St.
Wilmington, DE 19801
(302) 571-1195

J-M Trading Corp.
241 Frontage Rd., Ste. 47
Burr Ridge IL 60521
(708) 655-3305

Kel Instruments Co. Inc.
Box 2174
Vineyard Haven, MA 02568
(508) 693-7798

Kruger & Eckels Inc.
1406 E. Wilshire Ave.
Santa Ana, CA 92705
(714) 547-5165

Lesco Inc.
20005 Lake Rd.
Rocky River, OH 44116
(216) 333-9250

LaMotte Chemical Products
Co.
128 N. Queen St.
Chestertown, MD 21620
(301) 778-3100

Ted Mahr Supply Inc.
Box 150310
Cape Coral, FL 33915
(813) 574-2214

Metex Corp. Ltd.
12 Penn Dr. Unit 1
Weston, ON M9L 2A9
CANADA
(416) 749-1210

Midwest Trading,
GROmaster Div.
Box 384
St. Charles, IL 60174
(708) 742-1840

Myron L Co.
6231 Yarrow Dr. Ste. C
Carlsbad, CA 92009
(619) 438-2021

Troy Hygro-Systems Inc.
4096 Hwy. ES
East Troy, WI 53120
(414) 642-5928

Otis S. Twilley Seed Co.
Box 65
Trevose, PA 19047
(215) 639-8800

**Relative humidity
sensors and recorders**

Aerovent Fan & Eqt. Inc.
929 Terminal Rd.
Lansing, MI 48906
(517) 323-2930

Atomizing Systems Inc.
1 Hollywood Ave.
Ho-Ho-Kus, NJ 07423
(201) 447-1222

Ball Seed Co.
622 Town Rd.
West Chicago, IL 60185
(708) 231-3500

Egor-Hoppmann Corp.
Box 601
Chantilly, VA 22021
(703) 631-2700

Midwest Trading,
GROmaster Div.
Box 384
St. Charles, IL 60174
(708) 742-1840

Pacific Controlled
Environments
18301 Collier Box 26
Lake Elsinore, CA 92330
(714) 674-1556

Q-Com Inc.
2050 S. Grand Ave.
Santa Ana, CA 92705
(714) 540-6123

Remote Measurement
Systems
2633 Eastlake Ave. E.
Ste. 200
Seattle, WA 98102
(206) 328-2255

Rustak Instruments
Rt. 2 Middle Rd.
East Greenwich, RI 02818
(401) 884-6800

Spraying Systems Co.
N. Ave. at Schmale Road
Wheaton, IL 60188
(708) 665-5000

**Screening material for
greenhouses**

Ball Seed Co.
622 Town Rd.
West Chicago, IL 60185
(708) 231-3500

Hydro-Gardens Inc.
P.O. Box 9707
Colorado Springs, CO 80932
(800) 634-6362

Tredegar Film Products
1100 Boulders Pkwy.
Richmond, VA 23225
(804) 330-1222

Tredegar Film Products
P.O. Box 1762
Orange, CA 92666
(714) 978-2677

Soil moisture sensors

Midwest Trading,
  GROmaster Div.
Box 384
St. Charles, IL 60174
(708) 742-1840

Pacific Controlled
  Environments
18301 Collier
Box 26
Lake Elsinore, CA 92330
(714) 674-1556

Remote Measurement
  Systems
2633 Eastlake Ave. E.
Ste. 200
Seattle, WA 98102
(206) 328-2255

Seedburo Eqt. Co.
1022 W. Jackson Blvd.
Chicago, IL 60607
(312) 738-3700

H.B. Sherman Mfg. Co.
1450 Rowe Pkwy,
Poplar Bluff, MO 63901
(314) 785-5754

Spot Systems
5812 Machine Dr.
Huntington Beach, CA 92649
(714) 891-1115

Winland Electronics Inc.
418 S. 2nd St.
Box 473
Mankato, MN 56001
(507) 625-7231

**Thermometers and
temperature recorders**

Agro Dynamics
Building 3, Navy Yard
Brooklyn, NY 11205
(718) 596-3042

Ball Seed Co.
622 Town Rd.
West Chicago, IL 60185
(708) 231-3500

Biotherm Engineering
Box 6007
Petaluma, CA 94953
(707) 763-4444

DTR Co. Ltd.
Box 4101
Modesto, CA 95352
(209) 526-8691

Dansco Distributors
4442 27th Ave., W.
Seattle, WA 98199
(206) 282-7282

Delta T Sales
3576 Empleo, Unit 2
San Luis Obispo, CA 93401
(805) 546-8814

Growers Ghse. Supplies
Box 83
Vineland Station, ON L0R
  2E0
CANADA
(416) 562-7341

Harnois Industries Inc.
1044 Principale
St. Thomas (Joliette),
  PQ J0K 3L0
CANADA
(514) 756-1041

Heat-Timer Corp.
10 Dwight Pl.
Fairfield, NJ 07006
(201) 575-4004

Neogen Food Tech
620 Lesher Pl.
Lansing, MI 48912
(517) 372-9200

Pacific Controlled
  Environments
Box 26
Lake Elsinore, CA 92330
(714) 674-1556

Remote Measurement
  Systems
2633 Eastlake Ave. E.
Ste. 200
Seattle, WA 98102
(206) 328-2255

Rustak Instruments
Rt. 2, Middle Rd.
East Greenwich, RI 02818
(401) 884-6800

Ryan Instruments
Box 599
Redmond, WA 98073-0599
(206) 883-7926

Troy Hygro-Systems Inc.
4096 Hwy. ES
East Troy, WI 53120
(414) 642-5928

Winandy Ghse. Co. Inc.
2211 Peacock Rd.
Richmond, IN 47374
(317) 935-2111

Winland Electronics Inc.
418 S. 2nd St.
Box 473
Mankato, MN 56001
(507) 625-7231

Thermostats

AAA Associates Inc.
1445 S. 3rd St.
Niles, MI 49120
(616) 684-4073

Acme Engineering
  & Mfg. Corp.
Box 978
Muskogee, OK 74402
(918) 682-7791

Aerovent Fan & Eqt. Inc.
929 Terminal Rd.
Lansing, MI 48906
(517) 323-2930

American Coolair Corp.
Box 2300
Jacksonville, FL 32203
(904) 389-3646

Ball Seed Co.
622 Town Rd.
West Chicago, IL 60185
(708) 231-3500

Biotherm Engineering
Box 6007
Petaluma, CA 94953
(707) 763-4444

Clover Ghses. by Elliott Inc.
200 Weakley Lane Box 789
Smyrna, TN 37167
(615) 459-3863

Conley's Mfg. & Sales
4344 E. Mission Blvd.
Pomona, CA 91766
(714) 627-0981

Dansco Distributors
4442 27th Ave. W.
Seattle, WA 98199
(206) 282-7282

Delta T Sales
3576 Empleo, Unit 2
San Luis Obispo, CA 93401
(805) 546-8814

Harnois Industries Inc.
1044 Principale
St. Thomas (Joliette),
  PQ J0K 3L0
CANADA
(514) 756-1041

Albert J. Lauer Inc.
16700 Chippendale Ave. W
(Hwy. 3)
Rosemount, MN 55068
(612) 423-1651

Neogen Food Tech
620 Lesher Pl.
Lansing, MI 48912
(517) 372-9200

Pacific Coast Ghse. Mfg. Co.
8360 Industrial Ave.
Cotati, CA 94928
(707) 795-2164

Phonetics Inc.
101 State Rd.
Media, PA 19063
(215) 565-8520

Precision Growth Systems
3350 Scott Blvd., Bldg. 61
Santa Clara, CA 95054
(408) 727-6256

Q-Com Inc.
2050 S. Grand Ave.
Santa Ana, CA 92705
(714) 540-6123

Roundhouse by Groce
Box 1744
Cleveland, TX 77327
(713) 592-7474

X.S. Smith Inc.
Drawer X
Red Bank, NJ 07701
(201) 222-4600

Troy Hygro-Systems Inc.
4096 Hwy. ES
East Troy, WI 53120
(414) 642-5928

Vary Industries
Box 160
Grimsby, ON L3M 4N6
CANADA
(416) 945-9691

Wadsworth Control Systems
5541 Marshall St.
Arvada, CO 80002
(303) 424-4461

Winandy Ghse. Co. Inc.
2211 Peacock Rd.
Richmond, IN 47374
(317) 935-2111

Timers

AAA Associates Inc.
1445 S. 3rd St.
Niles, MI 49120
(616) 684-4073

Batrow Inc.
171 Short Beach Rd.
Short Beach, CT 06405
(203) 488-2578

Dansco Distributors
4442 27th Ave. W.
Seattle, WA 98199
(206) 282-7282

Davis Engineering
8217 Corbin Ave.
Canoga Park, CA 91306
(818) 993-0607

Midwest Trading,
GROmaster Div.
Box 384
St. Charles, IL 60174
(708) 742-1840

Pacific Coast Ghse. Mfg. Co.
8360 Industrial Ave.
Cotati, CA 94928
(707) 795-2164

Pacific Controlled
Environments
18301 Collier
Box 26
Lake Elsinore, CA 92330
(714) 674-1556

Precision Growth Systems
3350 Scott Blvd., Bldg. 61
Santa Clara, CA 95054
(408) 727-6256

Troy Hygro-Systems Inc.
4096 Hwy. ES
East Troy, WI 53120
(414) 642-5928

Vary Industries
Box 160
Grimsby, ON L3M 4N6
CANADA
(416) 945-9691

W.A. Westgate Co. Inc.
Box 445
Davis, CA 95617
(916) 753-2954

Winandy Ghse. Co. Inc.
2211 Peacock Rd.
Richmond, IN 47374
(317) 935-2111

SUPPLIERS OF BENEFICIAL INSECTS AND MITES

(Contact the individual supplier for information on prices and ability to supply the required beneficial insects/mites in the proper amounts. List sources: *Connecticut Greenhouse Newsletter* July 1990 and *Greenhouse Grower* June 1991.)

Ag Bio Chem Inc.
3 Fleetwood Ct.
Orinda, CA 94563
(415) 254-0789

Alternative
349 E. 86th St., Ste. 259
Indianapolis, IN 46240
(317) 823-0432

Applied Bionomics
P.O. Box 2637
Sidney, B.C.
CANADA, V8L 4C1
(604) 656-2123

ARBICO
P.O. Box 4247 CRB
Tucson, AZ 85738
(800) 505-2847

Associates Insectary
P.O. Box 969
Santa Paula, CA 93060
(805) 933-1301

Beneficial Bugs
Earl Nelson
P.O. Box 1627
Apopka, FL 32702-1627
(305) 886-2384

Beneficial Insectary
245 Oak Run Rd.
Oak Run, CA 96069
(916) 472-3715

Beneficial Insects Company
P.O. Box 556
Brownsville, CA 95919
(916) 675-2251

Better Yields Insects
13310 Riverside Dr. E.
Tecumseh, Ontario
CANADA N8N 1B2
(519) 735-0002

Bio-Con Systems
P.O. Box 377
Sunneymead, CA 92388
(714) 656-1712

Bio-Control Company
P.O. Box 247
Cedar Ridge, CA 95924
(916) 272-1997

Biogenesis Inc.
P.O. Box 36
Mathis, TX 78368
(512) 547-3259

Bio Insect Control
710 S. Columbia
Plainview, TX 79072
(806) 293-5861

Biotactics Inc.
7765 Lakeside Dr.
Riverside, CA 92509
(714) 685-7681

B.R. Supply Company
P.O. Box 845
Exeter, CA 93221
(209) 732-3422

California Green Lacewings
P.O. Box 2495
Merced, CA 95340
(209) 722-4985

Colorado Insectary
P.O Box 3266
Durango, CO 81302
(303) 247-5360

Fairfax Biological Labs Inc.
Clinton Corners, NY 12514
(914) 266-3705

Farmers Seed and Nursery
2207 E. Oakland Avenue
Bloomington, IL 61701
(309) 663-9551

Foothill Ag. Research Inc.
510 W. Chase Dr.
Corona, CA 91720
(714) 371-0120

Fountain's Sierra Bug Co.
P.O. Box 114
Rough and Ready, CA 95975
(916) 273-0513

Gerhart's
Avon Belden Rd.
North Ridgeville, OH
(216) 327-8056

Gothard Inc.
P.O. Box 7
Mersilla, NM 88046
(505) 552-9031

Gurney Seed and Nsy. Co.
Yankton, SD 57079
(605) 665-4451

Harmony Farm Supply
P.O Box 451
Grafton, CA 95444
(707) 832-9125

Henry Field's Seed and
Nursery Co.
Shenandoah, Iowa 51602
(712) 246-1888

Hydro-Gardens Inc.
P.O. Box 9797
Colorado Springs, CO 80932
(800) 634-6362

IPM Laboratories Inc.
Main Street
Locke, NY 13092
(315) 497-3129

Lakeland Nurseries Sales Inc.
340 Poplar Street
Hanover, PA 17331
(717) 637-5555

Mellinger's Nursery
2310 W. South Range Rd.
North Lima, OH 44452
(216) 549-9861

Natural Pest Controls
8864 Little Creek Dr.
Orangevale, CA 95662
(916) 726-0855

Nature's Control
P.O. Box 35
Medford, OR 97501
(513) 773-5927

Necessary Trading Company
P.O Box 603
New Castle, VA 24127
(703) 864-5103

The Nematode Farm Inc.
3335 Birch Street
Palo Alto, CA 94306
(415) 494-8630

Organic Pest Control
Naturally
P.O. Box 55267
Seattle, WA 98155
(206) 367-0707

Richard Owen Nusery
2300 E. Lincoln St.
Bloomington, IL 61701
(309) 663-9551

Pacific Tree Farms
4301 Lynwood Dr.
Chula Vista, CA 92010
(619) 422-2400

Peaceful Valley Farm Supply
11173 Peaceful Valley Rd.
Nevada City, CA 95959
(916) 265-3276

Reuter Laboratories
P.O. Box 346
Haymarket, VA 22069
(800) 368-2244

Richters
P.O. Box 26
Goodwood, ON L0C 1AO
CANADA
(416) 640-6677

Rincon-Vitova
Insectaries Inc.
P.O. Box 95
Oak View, AC 93022
(805) 643-5407

Rocky Mountain Insectary
P.O. Box 152
Palisade, CO 81526
(303) 245-0406

Spalding Laboratories
760 Printz Rd.
Arroyo Grande, CA 93420
(805) 489-5946

Unique Insect Control
P.O. Box 15376
Sacramento, CA 95851
(916) 967-7082

West Coast Ladybug Sales
P.O. Box 903
Gridley, CA 95948
(916) 534-0840

Wilk, Kitayama and Mead
9093 Troxel Rd.
Chico, CA 95928
(916) 895-8424

**MANUFACTURERS AND/
OR SUPPLIERS OF LOW-
AND ULTRA-LOW-VOL-
UME SPRAY APPLICA-
TION EQUIPMENT**

Aerosol formulations

Ball Seed Co.
622 Town Rd.
West Chicago, IL 60185
(708) 231-3500

Whitmire Research
Laboratories Inc.
3568 Tree Court
St. Louis, MO 63122-6620
(800) 325-3668

Appendix I

Electrostatic sprayers

Electrostatic Spraying
  Systems Inc.
1880 Commerce Rd. #107
Athens, GA
(404) 353-0695

Parker Hannifin Corporation
Advanced Spray Technology
17325 Euclid Ave.
Cleveland, OH 44112
(216) 531-3000

**Mechanical aerosol
generators, coldfoggers**

AgroDynamics
12 Elkins Rd.
East Brunswick, NJ 08816
(800) 872-2476

Ball Seed Co.
622 Town Rd.
West Chicago, IL 60185
(708) 231-3500

Dramm Intl. Inc.
P.O. Box 528
Manitowoc, WI 54220
(414) 684-0227

Electrostatic Spraying
  Systems Inc.
1880 Commerce Rd. #107
Athens, GA
(404) 353-0695

Flamingo Holland Inc.
USA: P.O. Box 387
San Luis Rey, CA 92068
(619) 967-1746
Canada: RR 7 Country Rd. 39
Strathroy, Ontario N7G 3H8
(519) 245-5665

Mykron Ghse. Technology
  Inc.
P.O. Box 282
Leamington, ON N8H 3W3
CANADA
(519) 326-7917

Rotary atomizers

London Fog Co.
505 Brimhall
P.O. Box 406
Long Lake, MN 55356
(612) 473-5366

Thermal pulse-jet foggers

AgroDynamics
12 Elkins Rd.
East Brunswick, NJ 08816
(800) 872-2476

Ball Seed Co.
622 Town Rd.
West Chicago, IL 60185
(708) 231-3500

Carmel Chemical Corporation
P.O. Box 406
Westfield, IN 46074
(317) 896-2531

Dramm Intl. Inc.
P.O. Box 528
Manitowoc, WI 54220
(414) 684-0227

292

Appendix II

Math and Conversion Tables

British Measurement System

Linear Measure

Unit	Equivalent values
Chain (engineers)	100 feet
Chain (surveyors)	66 feet
Fathom	72 inches or 6 feet
Foot	12 inches
Furlong	660 feet or 220 yards or 40 rods
Inch	0.0833 feet or 0.02778 yards
League	15,840 feet or 5,280 yards or 3 miles
Mil	0.001 inch
Mile	5,280 feet or 1,760 yards or 320 rods or 8 furlongs
Rod	$16^{1}/_{2}$ feet or $5^{1}/_{2}$ yards
Yard	36 inches or 3 feet

Square Measure

Unit	Equivalent values
Square acre	43,560 sq. feet or 4,840 sq. yards or 160 sq. rods
Square foot	144 sq. inches
Square mile	102,400 sq. rods or 640 acres
Square rod	$272^{1}/_{4}$ sq. feet or $30^{1}/_{4}$ sq. yards
Square yard	1,296 sq. inches or 9 sq. feet

Calculating Parts Per Million

In the horticulture industry when we speak of fertilizing crops, we generally think in terms of pounds of actual element per acre or per 1,000 square feet, or in terms of spoonfuls of a particular fertilizer per 3- or 5-gallon container. This approach is fine as long as we're dealing only with the application of dry fertilizers or nutrient elements. A serious problem develops, however, when we change over, either by choice or necessity, to liquid fertilizers. Suddenly we're faced with having to think in terms of parts per million (ppm). In addition, recommendations for applications of growth promoting compounds, chemical pinching and branching agents, growth retardants, and root promoting compounds are all given in parts per million. Accurate applications can be made only if the grower has a working knowledge of what is meant by parts per million and how to make some basic calculations. Once these are understood, it becomes relatively simple to work out those problems concerning ppm.

Parts per million refers to concentration of a material for any specific unit of weight (mass) or volume. For example, one blonde-haired person living in a city with 999,999 brown-haired people would represent .0001 percent of that population or 1 ppm; 1/4 ounce of lead shot mixed in with 249,999 ounces of steel shot would represent .0001 percent or 1 ppm. Although not scientifically accurate, growers use the rule of thumb that 1 ounce of a material in 100 gallons of water is equal to 75 ppm. For example, 1 ounce of pure nitrogen dissolved in 100 gallons of water is equivalent to 75 ppm nitrogen. Using this rule of thumb, calculating ppm becomes simple. Consider the following:

Problem 1: A grower wants to apply 225 ppm N to a crop of Japanese holly liners. The soluble fertilizer available is 20-20-20. How much 20-20-20 should he dissolve per 100 gallons of water?

Solution:
1 ounce per 100 gallons = 75 ppm
225 ppm ÷ 75 ppm = 3
3 ounces supplies 225 ppm
but
20-20-20 = 20 percent N
5 ounces 20-20-20 = 1 ounce N (20% of 5)
therefore
3 × 5 ounces 20-20-20 = 15 ounces of 20-20-20 needed

The grower should dissolve 15 ounces of 20-20-20 in 100 gallons of water to apply 225 ppm N. Of course the solution would also contain 225 ppm of P_2O_5 and 225 ppm of K_2O.

The procedure may also be reversed to determine the concentration of a fertilizer solution. Problem 2 illustrates this.

Problem 2: A 15-45-5 fertilizer is recommended for use at the rate of 3 pounds per 100 gallons of water. How much N, P and K are being applied by the solution?

Solution:
1 pound = 16 ounces
3 pounds = 48 ounces
48 ounces $\times$ .15 N $\times$ 75 ppm/ounce = 540 ppm N
48 ounces $\times$ .45 P_2O_5 $\times$ 75 ppm/ounce = 1,620 ppm P_2O_5
48 ounces $\times$ .05 K_2O $\times$ 75 ppm/ounce = 180 ppm K_2O

Problem 3: A grower would like to fertilize his crop of photinia liners with 225 ppm N using a 20-20-20 fertilizer and he would like to apply it through a proportioner that has a dilution ration of 1:15. How much 20-20-20 should be dissolved per gallon of concentrated stock solution?

Solution:
15 ounces 20-20-20 per 100 gallons water = 225 ppm N
1:15 proportioner = 1 gallon stock solution per 15 gallons water
Total volume = 16 gallons
100 gallons ÷ 16 gallons = 6.25
15 ounces ÷ 6.25 = 2.4 ounces per gallon

To apply 225 ppm N through the 1:15 proportioner, dissolve 2.4 ounces of 20-20-20 in each gallon of stock solution.

Problem 4: A grower would like to spray a particular growth regulator on his azalea crop at a concentration of 2,000 ppm. The growth regulator as purchased from the supplier contains 18.5% active ingredients (ai). How much growth regulator should he use per gallon of spray?

Solution:
2,000 ppm = .2 percent = .002
1 gallon = 128 ounces
128 ounces $\times$ .002 = 0.256 ounces
Divide 100 percent purity by actual rates of ai.
100 percent ÷ 18.5 percent = 5.4
5.4 ounces material = 1 ounce ai
5.4 $\times$ 0.256 = 1.38 ounces per gallon = 2,000 ppm

Conversion Factors

Multiply	By	To obtain
Acres	43,560	Square feet
Acres	.4047	Hectares
Acres	4,047	Square meters
Acres	.001562	Square miles
Acres	160	Square rods
Acres	4,840	Square yards
Acre feet	12	Acre inches
Acre feet	43,560	Cubic feet
Acre feet	325,872	Gallons
Acre inches	3,630	Cubic feet
Acre inches	6,272,640	Cubic inches
Acre inches	27,154	Gallons
Board feet	144	Cubic inches
Bushels	1.244	Cubic feet
Bushels	2,150	Cubic inches
Bushels	.03524	Cubic meters
Bushels	.04545	Cubic yards
Bushels	35.238	Liters
Bushels	4	Pecks
Bushels	64	Pints (dry)
Bushels	32	Quarts (dry)
Centigrams	.01	Grams
Centiliters	.01	Liters
Centimeters	.03281	Feet
Centimeters	.3937	Inches
Centimeters	.01	Meters
Centimeters	393.7	Mils
Centimeters per second	1.969	Feet per minute
Centimeters per second	.03281	Feet per second
Centimeters per second	.036	Kilometers per hour
Centimeters per second	.6	Meters per minute
Centimeters per second	.02237	Miles per hour
Centimeters per second	.0003728	Miles per minute
Cords	128	Cubic feet
Cubic centimeters	.0000353	Cubic feet
Cubic centimeters	.06102	Cubic inches
Cubic centimeters	.00001	Cubic meters
Cubic centimeters	.000001308	Cubic yards
Cubic centimeters	.0002642	Gallons
Cubic centimeters	.001	Liters
Cubic centimeters	1	Milliliters
Cubic centimeters	.03382	Ounces (fluid)

Multiply	By	To obtain
Cubic centimeters	.002113	Pints (fluid)
Cubic centimeters	.001057	Quarts (fluid)
Cubic feet	.8	Bushels
Cubic feet	.0078	Cords
Cubic feet	28,316.84	Cubic centimeters
Cubic feet	1,728	Cubic inches
Cubic feet	.02832	Cubic meters
Cubic feet	.03704	Cubic yards
Cubic feet	7.481	Gallons
Cubic feet	28.32	Liters
Cubic feet	51.42	Pints (dry)
Cubic feet	59.84	Pints (fluid)
Cubic feet	25.71	Quarts (dry)
Cubic feet	29.92	Quarts (fluid)
Cubic feet per minute	472	Cubic centimeters per second
Cubic feet per minute	448.8	Gallons per hour
Cubic feet per minute	.1247	Gallons per second
Cubic feet per minute	1,698.74	Liters per hour
Cubic feet per minute	.4720	Liters per second
Cubic feet per minute	62.4	Pounds of water per minute
Cubic inches	.000465	Bushels
Cubic inches	16.387	Cubic centimeters
Cubic inches	.0005787	Cubic feet
Cubic inches	.00001639	Cubic meters
Cubic inches	.00002143	Cubic yards
Cubic inches	.004329	Gallons
Cubic inches	.01639	Liters
Cubic inches	.5541	Ounces (fluid)
Cubic inches	.02976	Pints (dry)
Cubic inches	.03463	Pints (fluid)
Cubic inches	.01488	Quarts (dry)
Cubic inches	.01732	Quarts (fluid)
Cubic meters	1,000,000	Cubic centimeters
Cubic meters	35.31	Cubic feet
Cubic meters	61,023	Cubic inches
Cubic meters	1.308	Cubic yards
Cubic meters	264.2	Gallons
Cubic meters	1,000	Liters
Cubic meters	2,113	Pints (fluid)
Cubic meters	1,057	Quarts (fluid)
Cubic yards	22	Bushels
Cubic yards	764,600	Cubic centimeters
Cubic yards	27	Cubic feet
Cubic yards	46,656	Cubic inches
Cubic yards	.7646	Cubic meters
Cubic yards	202	Gallons

Multiply	By	To obtain
Cubic yards	764.6	Liters
Cubic yards	1,616	Pints (fluid)
Cubic yards	807.9	Quarts (fluid)
Cubic yards per minute	.45	Cubic feet per second
Cubic yards per minute	3.367	Gallons per second
Cubic yards per minute	12.74	Liters per second
Cups (dry)	.5	Pints (dry)
Cups (dry)	.25	Quarts (dry)
Cups (dry)	16	Tablespoons (dry)
Cups (dry)	48	Teaspoons (dry)
Cups (fluid)	.5	Pints (fluid)
Cups (fluid)	.25	Quarts (fluid)
Cups (fluid)	16	Tablespoons (fluid)
Cups (fluid)	48	Teaspoons (fluid)
Days	24	Hours
Days	1,440	Minutes
Days	86,400	Seconds
Decigrams	.1	Grams
Deciliters	.1	Liters
Decimeters	10	Meters
Dekagrams	10	Grams
Dekaliters	10	Liters
Dekameters	10	Meters
Drams	27.343	Grains
Drams	1.772	Grams
Drams	.0625	Ounces
Fathoms	6	Feet
Feet	30.48	Centimeters
Feet	12	Inches
Feet	.3048	Meters
Feet	.0606	Rods
Feet	.333	Yards
Feet of water	.02950	Atmospheres
Feet of water	.8826	Inches of mercury
Feet of water	304.8	Kilograms per square meter
Feet of water	62.43	Pounds per square foot
Feet of water	.4335	Pounds per square inch
Feet per minute	.5080	Centimeters per second
Feet per minute	.01667	Feet per second
Feet per minute	.01829	Kilometers per hour
Feet per minute	.3048	Meters per minute
Feet per minute	.01136	Miles per hour
Feet per second	30.48	Centimeters per second
Feet per second	1.097	Kilometers per hour
Feet per second	.5921	Knots per hour
Feet per second	18.29	Meters per minute

Multiply	By	To obtain
Feet per second	.6818	Miles per hour
Feet per second	.01136	Miles per minute
Feet of rise per 100 feet	1	Percent grade
Footcandles	10.76	Lux
Furlongs	4	Rods
Gallons	3,785	Cubic centimeters
Gallons	.1337	Cubic feet
Gallons	231	Cubic inches
Gallons	.003785	Cubic meters
Gallons	.004951	Cubic yards
Gallons	3.785	Liters
Gallons	128	Ounces (fluid)
Gallons	8	Pints (fluid)
Gallons	4	Quarts (fluid)
Gallons of water	8.3453	Pounds of water
Gallons per minute	.134	Cubic feet per minute
Gallons per minute	.002228	Cubic feet per second
Gallons per minute	.06308	Liters per second
Gills	2	Cups (fluid)
Gills	.1183	Liters
Gills	4	Ounces (fluid)
Gills	.25	Pints (fluid)
Grains (troy)	1	Grains (avoir.)
Grains (troy)	.0648	Grams
Grains (troy)	.4167	Pennyweights (troy)
Grams	15.43	Grains (troy)
Grams	.001	Kilograms
Grams	1000	Milligrams
Grams	.03527	Ounces (avoir.)
Grams	.03215	Ounces (troy)
Grams	.002205	Pounds
Grams per liter	1,000	Parts per million
Grams per liter	.1336	Ounces per gallon
Grams per liter	.0334	Ounces per quart
Hectares	2.471	Acres
Hectares	107,000	Square feet
Hectograms	100	Grams
Hectoliters	100	Liters
Hectometers	100	Meters
Horsepower	42.44	BTUs per minute
Horsepower	.7457	Kilowatts
Horsepower	745.7	Watts
Horsepower (boiler)	33,520	BTUs per hour
Horsepower (boiler)	9.804	Kilowatts
Horsepower (boiler)	9,804	Watts
Horsepower hours	2,547	BTUs

Multiply	By	To obtain
Horsepower hours	.7457	Kilowatt hours
Hours	60	Minutes
Hours	3,600	Seconds
Inches	2.540	Centimeters
Inches	.08333	Feet
Inches	.0254	Meters
Inches	.02778	Yards
Inches of water	.002458	Atmospheres
Inches of water	.07355	Inches of mercury
Inches of water	25.4	Kilograms per square meter
Inches of water	.5781	Ounces per square inch
Inches of water	5.204	Pounds per square foot
Inches of water	.03613	Pounds per square inch
Kilograms	1,000	Grams
Kilograms	35.27	Ounces (avoir.)
Kilograms	2.2046	Pounds
Kilograms	.001102	Tons (short)
Kilogram meters	7.233	Foot pounds
Kilograms per hectare	.8929	Pounds per acre
Kilograms per cubic meter	.001	Grams per cubic centimeter
Kilograms per cubic meter	.06243	Pounds per cubic foot
Kilograms per cubic meter	.00003613	Pounds per cubic inch
Kilograms per square meter	.001422	Pounds per square inch
Kiloliters	1,000	Liters
Kilometers	100,000	Centimeters
Kilometers	3,281	Feet
Kilometers	1,000	Meters
Kilometers	.6214	Miles
Kilometers	1,093.6	Yards
Kilometers per hour	27.78	Centimeters per second
Kilometers per hour	54.68	Feet per minute
Kilometers per hour	.9114	Feet per second
Kilometers per hour	.5396	Knots per hour
Kilometers per hour	16.67	Meters per minute
Kilometers per hour	.6214	Miles per hour
Kilowatts	56.92	BTUs per minute
Kilowatts	1.341	Horsepower
Kilowatts	1,000	Watts
Kilowatt hours	3.415	BTUs
Kilowatt hours	1.341	Horsepower hours
Knots	6,080	Feet
Knots	1.853	Kilometers
Knots	1.152	Miles
Knots	2,027	Yards
Knots per hour	51.48	Centimeters per second
Knots per hour	1.689	Feet per second

Multiply	By	To obtain
Knots per hour	1.853	Kilometers per hour
Knots per hour	1.152	Miles per hour
Liters	1,000	Cubic centimeters
Liters	.03531	Cubic feet
Liters	61.02	Cubic inches
Liters	.001	Cubic meters
Liters	.001308	Cubic yards
Liters	.2642	Gallons
Liters	1,000	Milliliters
Liters	2.113	Pints (fluid)
Liters	1.057	Quarts (fluid)
Lux	.0929	Footcandles
Meters	100	Centimeters
Meters	3.2808	Feet
Meters	39.37	Inches
Meters	.001	Kilometers
Meters	1,000,000	Microns
Meters	1,000	Millimeters
Meters	1.0936	Yards
Meters per minute	1.667	Centimeters
Meters per minute	3.281	Feet per minute
Meters per minute	.5468	Feet per second
Meters per minute	.06	Kilometers per hour
Meters per minute	.03728	Miles per hour
Meters per second	1,968	Feet per minute
Meters per second	32.84	Feet per second
Meters per second	3.0	Kilometers per hour
Meters per second	.06	Kilometers per minute
Meters per second	2.237	Miles per hour
Meters per second	.03728	Miles per minute
Microns	.0001	Centimeters
Microns	.00003937	Inches
Microns	.000001	Meters
Microns	.001	Millimeters
Mils	.00254	Centimeters
Mils	.001	Inches
Miles	160,900	Centimeters
Miles	5,280	Feet
Miles	63,360	Inches
Miles	1.6093	Kilometers
Miles	1,609.3	Meters
Miles	320	Rods
Miles	1,760	Yards
Miles per hour	44.7	Centimeters per second
Miles per hour	88	Feet per minute
Miles per hour	1.467	Feet per second

Multiply	By	To obtain
Miles per hour	1.6093	Kilometers per hour
Miles per hour	.8684	Knots per hour
Miles per hour	26.82	Meters per minute
Milligrams	.001	Grams
Milligrams	.000001	Kilograms
Milligrams per liter	1	Parts per million
Milligrams per liter	.0001	Percent
Milliliters	1	Cubic centimeters
Milliliters	.001	Liters
Millimeters	1	Centimeters
Millimeters	.03937	Inches
Millimeters	.001	Meters
Millimeters	39.37	Mils
Months	30.42	Days
Months	730	Hours
Months	43,800	Minutes
Months	2,628,000	Seconds
Ounces (avoir.)	16	Drams
Ounces (avoir.)	437.5	Grains
Ounces (avoir.)	28.35	Grams
Ounces (avoir.)	.9115	Ounces (troy)
Ounces (avoir.)	3	Tablespoons (dry)
Ounces (avoir.)	9	Teaspoons (dry)
Ounces (fluid)	1.80	Cubic inches
Ounces (fluid)	.0078125	Gallons
Ounces (fluid)	.02957	Liters
Ounces (fluid)	29.57	Milliliters
Ounces (fluid)	2	Tablespoons (fluid)
Ounces (fluid)	6	Teaspoons (fluid)
Ounces (troy)	480	Grains (troy)
Ounces (troy)	31.10	Grams
Ounces (troy)	1.097	Ounces (avoir.)
Ounces (troy)	20	Pennyweights (troy)
Ounces (troy)	.8333	Pounds (troy)
Ounces per gallon	7.812	Milliliters per liter
Ounces per square inch	.625	Pounds per square inch
Parts per million	.001	Grams per liter
Parts per million	1	Milligrams per kilogram
Parts per million	1	Milligrams per liter
Parts per million	.013	Ounces per 100 gallons
Parts per million	.0001	Percent
Parts per million	.0083	Pounds per 1000 gallons
Pennyweights (troy)	24	Grains (troy)
Pennyweights (troy)	1.555	Grams
Pennyweights (troy)	.05	Ounces (troy)
Percent	10	Grams per kilogram

Multiply	By	To obtain
Percent	10	Grams per liter
Percent	1.33	Ounces by weight per gallon of water
Percent	10,000	Parts per million
Percent	8.34	Pounds per 100 gallons of water
Pints (dry)	.015625	Bushels
Pints (dry)	.0194	Cubic feet
Pints (dry)	33.6	Cubic inches
Pints (dry)	.0625	Pecks
Pints (dry)	.5	Quarts (dry)
Pints (fluid)	473.167	Cubic centimeters
Pints (fluid)	.0167	Cubic feet
Pints (fluid)	28.875	Cubic inches
Pints (fluid)	.125	Gallons
Pints (fluid)	.4732	Liters
Pints (fluid)	16	Ounces (fluid)
Pints (fluid)	.5	Quarts (fluid)
Pounds	256	Drams
Pounds	7,000	Grains
Pounds	453.594	Grams
Pounds	.453494	Kilograms
Pounds	16	Ounces
Pounds	14.583	Ounces (troy)
Pounds	1,215	Pounds (troy)
Pounds	.0005	Tons (short)
Pounds of water	.01602	Cubic feet
Pounds of water	27.68	Cubic inches
Pounds of water	.1198	Gallons
Pounds per acre	1.12	Kilograms per hectare
Pounds per cubic foot	.01602	Cubic feet
Pounds per cubic foot	16.02	Kilograms per cubic meter
Pounds per cubic foot	.005787	Pounds per cubic inch
Pounds per cubic inch	27.68	Pounds per cubic centimeter
Pounds per cubic inch	27,680	Kilograms per cubic meter
Pounds per cubic inch	1,728	Pounds per cubic foot
Pounds per foot	1,488	Kilograms per meter
Pounds per inch	178.6	Grams per centimeter
Pounds per square foot	.01602	Cubic feet of water
Pounds per square foot	4.882	Kilograms per square meter
Pounds per square foot	.006994	Pounds per square inch
Pounds per square inch	.06804	Atmospheres
Pounds per square inch	2.307	Cubic feet of water
Pounds per square inch	2.036	Inches of mercury
Pounds per square inch	.070307	Kilograms per square centimeter
Pounds per square inch	703.1	Kilograms per square meter
Pounds per square inch	144	Pounds per square foot

Multiply	By	To obtain
Quarts (dry)	.03125	Bushels
Quarts (dry)	.0389	Cubic feet
Quarts (dry)	67.20	Cubic inches
Quarts (dry)	.125	Pecks
Quarts (dry)	2	Pints (dry)
Quarts (fluid)	.0334	Cubic feet
Quarts (fluid)	57.75	Cubic inches
Quarts (fluid)	.25	Gallons
Quarts (fluid)	.9463	Liters
Quarts (fluid)	946.3	Milliliters
Quarts (fluid)	32	Ounces (fluid)
Quarts (fulid)	2	Pints (fluid)
Rods	16.5	Feet
Rods	198	Inches
Rods	5.029	Meters
Rods	5.5	Yards
Square centimeters	.00107	Square feet
Square centimeters	.1550	Square inches
Square centimeters	.0001	Square meters
Square centimeters	100	Square millimeters
Square feet	.00002296	Acres
Square feet	929	Square centimeters
Square feet	144	Square inches
Square feet	.0929	Square meters
Square feet	.0000000357	Square miles
Square feet	.111	Square yards
Square inches	6.452	Square centimeters
Square inches	.006944	Square feet
Square inches	645.163	Square millimeters
Square kilometers	247.1	Acres
Square kilometers	10,764,961	Square feet
Square kilometers	1,000,000	Square meters
Square kilometers	.3861	Square miles
Square kilometers	1,196,107	Square yards
Square meters	.000247	Acres
Square meters	10.764	Square feet
Square meters	.0000003861	Square miles
Square meters	1.196	Square yards
Square miles	640	Acres
Square miles	27,878,400	Square feet
Square miles	2.59	Square kilometers
Square miles	102,400	Square rods
Square miles	3,097,600	Square yards
Square millimeters	.01	Square centimeters
Square millimeters	.00155	Square inches

Multiply	By	To obtain
Square millimeters	.000001	Square meters
Square yards	.0002066	Acres
Square yards	9	Square feet
Square yards	1,296	Square inches
Square yards	.8361	Square meters
Square yards	.000000322	Square miles
Tablespoons (dry)	.0625	Cups (dry)
Tablespoons (dry)	.333	Ounces (dry)
Tablespoons (dry)	3	Teaspoons (dry)
Tablespoons (fluid)	.0625	Cups (fluid)
Tablespoons (fluid)	15	Milliliters
Tablespoons (fluid)	.5	Ounces (fluid)
Teaspoons (dry)	.111	Ounces (dry)
Teaspoons (dry)	.333	Tablespoons (dry)
Teaspoons (fluid)	.0208	Cups (fluid)
Teaspoons (fluid)	5	Milliliters
Teaspoons (fluid)	.1666	Ounces (fluid)
Temperature (°C) +17.8	1.8	Temperature °F
Temperature (°F) -32	.55	Temperature °C
Tons (long)	1,016	Kilograms
Tons (long)	2,240	Pounds
Tons (long)	1.016	Tons (long)
Tons (long)	1.1199	Tons (short)
Tons (metric)	1,000	Kilograms
Tons (metric)	2,205	Pounds
Tons (metric)	.9843	Tons (long)
Tons (metric)	1.1023	Tons (short)
Tons (short)	907.2	Kilograms
Tons (short)	2,000	Pounds
Tons (short)	.8929	Tons (long)
Tons (short)	.9072	Tons (metric)
Watts	.05692	BTUs per minute
Watts	.001341	Horsepower
Watts	.001	Kilowatts
Weeks	168	Hours
Weeks	10,080	Minutes
Weeks	604,800	Seconds
Yards	.009144	Centimeters
Yards	3	Feet
Yards	36	Inches
Yards	.9144	Meters
Yards	.000568	Miles
Yards	.01818	Rods
Years (common)	365	Days
Years (common)	8.760	Hours
Years (leap)	366	Days

Fraction, Decimal and Millimeter Equivalents

Fraction of an inch	Inches in decimal	Millimeter equivalent	Fraction of an inch	Inches in decimal	Millimeter equivalent
1/64	.015625	0.397	17/32	.53125	13.494
1/32	.03125	0.794	35/64	.546875	13.891
3/64	.046875	1.191	9/16	.5625	14.288
1/16	.0625	1.588	37/64	.578125	14.684
5/64	.078125	1.984	19/32	.59375	15.081
3/32	.09375	2.381	39/64	.609375	15.478
7/64	.109375	2.778	5/8	.6250	15.875
1/8	.1250	3.572	41/64	.640625	16.272
5/32	.15625	3.969	21/32	.65625	16.669
11/64	.171875	4.366	43/64	.671875	17.066
3/16	.1875	4.762	11/16	.6875	17.462
13/64	.203125	5.159	45/64	.703125	17.859
7/32	.21875	5.556	23/32	.71875	18.256
15/64	.234375	5.593	47/64	.734375	18.653
1/4	.2500	6.350	3/4	.7500	19.050
17/64	.265625	6.747	49/64	.765625	19.447
9/32	.28125	7.144	25/32	.78125	19.844
19/64	.296875	7.541	51/64	.796875	20.241
5/16	.3125	7.938	13/16	.8125	20.638
21/64	.328125	8.334	53/64	.828125	21.034
11/32	.34375	8.731	27/32	.84375	21.431
23/64	.359375	9.128	55/64	.859375	21.828
3/8	.3750	9.525	7/8	.8750	22.225
25/64	.390625	9.922	57/64	.890625	22.622
13/32	.40625	10.319	29/32	.90625	23.019
27/64	.421875	10.716	59/64	.921875	23.416
7/16	.4375	11.112	15/16	.9375	23.812
29/64	.453125	11.509	61/64	.953125	24.209
15/32	.46875	11.906	31/32	.96875	24.606
31/64	.484375	12.303	63/64	.984375	25.003
1/2	.5000	12.700	1	1.000	25.400
33/64	.515625	13.097			

The Metric System

The metric system is based on three basic units of measure: the meter, liter and gram. Multiples of fractions of the three basic units are denoted by adding specific prefixes to the basic unit:

Metric prefix	Exponential multiplier	Numeric multiplier	Increment of basic unit
—	10	1	One
Deka-	10	10	Ten
Hecto-	10	100	One hundred
Kilo-	10	1,000	One thousand
Mega-	10	1,000,000	One million
Giga-	10	1,000,000,000	One billion
Tera-	10	1,000,000,000,000	One trillion
—	10	1	One
Deci-	10	.1	One tenth
Centi-	10	.01	One hundreth
Milli-	10	.001	One thousandth
Micro-*	10	.000001	One millionth
Nano-	10	.000000001	One billionth
Pico-	10	.000000000001	One trillionth

*An exception to the rule of metric prefixes is that one millionth of a meter is referred to as a micron rather than a micrometer.

Linear measure (basic unit is the meter)

Unit	Abbreviation	Equivalent values
Kilometer	km	1,000 meters
Hectometer	hm	100 meters
Decameter	dkm	10 meters
Meter	m	1,000 millimeters
Decimeter	dm	.1 meters
Centimeter	cm	.01 meters
Millimeter	mm	.001 meters
Micron		.000001 meters
Millimicron		.000000001 meters

Square measure (basic unit is the meter)

Unit	Equivalent value
Kilometer	1,000,000 square meters
Meter	10,000 square centimeters
Centimeter	100 square millimeters
Millimeter	.01 square centimeters

Determining Number of Plants Required Per Acre

1. $$\frac{\text{Plants per acre}}{\text{Row spacing (in feet)} \times \text{Plant spacing (in feet)}} = 43{,}560$$

OR

2. $$\frac{\text{Plants per acre}}{\text{Row spacing (in inches)} \times \text{Plant spacing (in inches)}} = 43{,}560 \times 144$$

Distance between rows (in feet)	Distance between plants in the row (in inches)				
	12	24	36	48	60
1	43,560	21,780	14,520	10,890	8,712
2	21,780	10,890	7,260	5,445	4,356
3	14,520	7,260	4,840	3,630	2,904
4	10,890	5,445	3,630	2,722	2,178
5	8,712	4,356	2,904	2,178	1,742
6	7,260	3,630	2,420	1,815	1,452
7	6,223	3,111	2,074	1,556	1,245
8	5,445	2,722	1,815	1,361	1,089
9	4,840	2,420	1,613	1,210	968
10	4,356	2,178	1,452	1,089	871
12	3,620	1,810	1,210	907	726
14	3,111	1,555	1,037	778	622
16	2,722	1,361	907	680	544
18	2,420	1,210	807	605	484
20	2,178	1,089	726	544	436

Conversions and "Rules of Thumb" for Pesticide Application

1) Surface

 1 square inch = 6.5 square centimeters
 1 square foot = 929 square centimeters = 0.0929 square meters
 1 square yard = 0.84 square meters
 43,560 square feet = 1 acre
 2.5 acres = 1 hectare = 10,000 square meters
 1 quart per 100 square feet = 100 gallons per acre

2) Dry Weight

 1 ounce = 28.35 grams
 1 pound = 454 grams = 16 ounces
 1 pound of most wettable powders per 100 gallons is approx. 1 tablespoon
  per gallon
 1 tablespoon = 3 teaspoons
 1 ounce active per 100 gallons = 75 ppm
 1 ppm = 1 mg. per 100 gms. = 0.001 ml. per liter
 1 gram per 100 square feet = 1 pound per acre

3) Liquid

 1 ounce = 29.6 mls. = 2 tablespoons = 6 teaspoons
 8 ounces = 1 cup
 2 cups = 1 pint
 2 pints = 1 quart
 4 quarts = 1 gallon
 10 liters = 2.64 gallons
 1 gallon = 128 ounces = 3800 ml. = 8.34 pounds of water
 1 gallon of concentrate per 100 gallons of spray = $2^1/_2$ tablespoons per gallon
 1 quart per 100 gallons = $^5/_8$ tablespoons per gallon
 1 pint = 1 pound of water

Pesticide Dilutions

The recommended rate for many commercial products is given in either gallons or pounds of product per 100 gallons of water. The following table can be used to determine the amount of commercial product to use when mixing less than 100 gallons of material.

Product needed to mix

100 gallons	25 gallons	20 gallons	15 gallons	10 gallons	5 gallons	1 gallon

Liquid formulations

100 gallons	25 gallons	20 gallons	15 gallons	10 gallons	5 gallons	1 gallon
2 gals.	64 oz.	$51^3/_{16}$ oz.	$38^1/_2$ oz.	$25^1/_2$ oz.	$12^7/_8$ oz.	$2^1/_2$ oz.
1 gal.	32 oz.	$25^9/_{16}$ oz.	$19^3/_{16}$ oz.	$12^3/_4$ oz.	$6^1/_2$ oz.	$1^1/_4$ oz.
2 qts.	16 oz.	$12^{13}/_{16}$ oz.	$9^9/_{16}$ oz.	$6^3/_8$ oz.	$3^1/_4$ oz.	$^5/_8$ oz.
1 qt.	8 oz.	$6^3/_8$ oz.	$4^{13}/_{16}$ oz.	$3^3/_{16}$ oz.	$1^9/_{16}$ oz.	$^5/_{16}$ oz.
$1^1/_2$ pints	6 oz.	$4^{13}/_{16}$ oz.	$3^9/_{16}$ oz.	$2^3/_8$ oz.	$1^1/_4$ oz.	$^1/_4$ oz.
1 pint	4 oz.	$3^3/_{16}$ oz.	$2^3/_8$ oz.	$1^9/_{16}$ oz.	$^7/_8$ oz.	$^3/_{16}$ oz.
8 oz.	2 oz.	$1^9/_{16}$ oz.	$1^3/_{16}$ oz.	$^{13}/_{16}$ oz.	$^7/_{16}$ oz.	$^1/_2$ tsp.
4 oz.	1 oz.	$^{13}/_{16}$ oz.	$^9/_{16}$ oz.	$^3/_8$ oz.	$^1/_4$ oz.	$^1/_4$ tsp.

Solid formulations

100 gallons	25 gallons	20 gallons	15 gallons	10 gallons	5 gallons	1 gallon
5 lbs.	20 oz.	16 oz.	12 oz.	8 oz.	4 oz.	$4^4/_5$ tsp.
4 lbs.	16 oz.	$12^{13}/_{16}$ oz.	$9^9/_{16}$ oz.	$6^3/_8$ oz.	$3^1/_4$ oz.	$3^4/_5$ tsp.
3 lbs.	12 oz.	$9^9/_{16}$ oz.	$7^3/_{16}$ oz.	$4^{13}/_{16}$ oz.	$2^3/_8$ oz.	$2^2/_5$ tsp.
2 lbs.	8 oz.	$6^3/_8$ oz.	$4^3/_8$ oz.	$3^3/_{16}$ oz.	$1^3/_4$ oz.	2 tsp.
1 lb.	4 oz.	$3^3/_{16}$ oz.	$2^3/_8$ oz.	$1^9/_{16}$ oz.	$^7/_8$ oz.	1 tsp.
8 oz.	2 oz.	$1^9/_{16}$ oz.	$1^{13}/_{16}$ oz.	$^{13}/_{16}$ oz.	$^3/_8$ oz.	$^1/_2$ tsp.
4 oz.	1 oz.	$^{13}/_{16}$ oz.	$^9/_{16}$ oz.	$^3/_8$ oz.	$^3/_{16}$ oz.	$^1/_4$ tsp.

Pots: Volume and Measurement

Pot diameter	Pot height	Trade designation	Actual dimensions (diameter x height)	Pot volume in cubic inches
6"	5"	1 gal. std.	6½" x 6"	140
6" tub	—	6" tub	6½" x 5"	—
8"	7"	2 gal.	8" x 7"	350
9"	8"	—	—	512
10"	9"	3 gal.	10" x 9½"	711
11"	10"	—	—	950
12"	—	4 gal.	11" x 10 1/2"	—
13"	11"	—	—	1,463
14"	—	7 gal.	13½" x 12"	—
17"	14"	10 gal.	17" x 15"	3,178

Cubic Inches to Cubic Feet

Cubic inches	Cubic feet
1,728 =	1
1,296 =	3/4
864 =	1/2
519 =	1/3
432 =	1/4

Potted Plants Required per 100 Square Feet
(Various spacings on center*)

Pot spacing	Plants per 100 sq. ft.	Pot spacing	Plants per 100 sq. ft.
4 x 4	900	18 x 18	45
6 x 6	400	24 x 24	25
8 x 8	225	30 x 30	16
9 x 9	178	36 x 36	11.11
10 x 10*	144*	48 x 48	6.25
12 x 12	100	72 x 72	2.78

*Example: If you measure and mark every 10 inches and place and center a plant right on top of each spot, it would take 144 plants to fill 100 square feet.

Application Rates of Media or Surface-Applied Fertilizers or Pesticides

Rate per acre	Rate per 1,000 sq. ft.	Rate per 100 sq. ft.
100 pounds	2 pounds 4 ounces	$3^1/_2$ ounces
200 pounds	4 pounds 8 ounces	$7^1/_4$ ounces
300 pounds	6 pounds 14 ounces	11 ounces
400 pounds	9 pounds	$14^1/_2$ ounces
500 pounds	11 pounds 8 ounces	1 pound 2 ounces
600 pounds	13 pounds 12 ounces	1 pound 6 ounces
700 pounds	16 pounds	1 pound 9 ounces
800 pounds	18 pounds	1 pound 13 ounces
900 pounds	20 pounds 9 ounces	2 pounds 1 ounces
1,000 pounds	23 pounds	2 pounds 5 ounces
2,000 pounds	46 pounds	4 pounds 9 ounces
1 gallon	3 ounces (fluid)	$^1/_3$ ounce (fluid)
5 gallons	$14^1/_2$ ounces (fluid)	$1^1/_2$ ounces (fluid)
100 gallons	$2^1/_3$ gallons (fluid)	$29^1/_3$ ounces (fluid)

Rate per acre	Rate per 1,000 sq. ft.	Rate per 100 sq. ft.
218 pounds	5 pounds	8 ounces.
436 pounds	10 pounds	1 pound
2,178 pounds	50 pounds	5 pounds
4,356 pounds	100 pounds	10 pounds
$43^1/_2$ gallons	1 gallon	13 ounces (fluid)
5 gallons $3^1/_2$ pints	1 pint	$1^1/_2$ ounces (fluid)
10 gallons $3^1/_2$ quarts	1 quart	3 ounces (fluid)

Weight (Mass) Conversions

Unit	Equivalent values
Dram (avoirdupois)	27.34 grains
Dram (troy)	60 grains
Ounce (avoirdupois)	437.5 grains
Ounce (troy)	480 grains
Pennyweight (avoirdupois)	0.877 dram
Pennyweight (troy)	0.4 dram
Pound (avoirdupois)	7,000 grains or 16 ounces
Pound (troy)	5,760 grains or 12 ounces
Ton (long)	2,240 pounds
Ton (metric)	2,205 pounds
Ton (short)	2,000 pounds

Determining Amount of Soluble Fertilizer or Pesticide Suspension for Proportioner Stock Solutions to Deliver 100 ppm

Ounces of fertilizer to dissolve per 1 gallon of concentrate to deliver 100 ppm

% N, P_2O_5 or K_2O in fertilizer	Proportioner dilution ratio					
	1:15	1:100	1:128	1:150	1:200	1:300
5	4.05	27.0	34.6	40.5	54.0	81.0
7	2.89	19.3	24.7	28.9	38.6	57.9
10	2.02	13.5	17.3	20.3	27.0	40.5
15	1.35	9.0	11.5	13.5	18.0	27.0
20	1.01	6.8	8.6	10.1	13.6	20.3
21	0.96	6.4	8.2	9.6	12.9	19.3
25	0.81	5.4	6.9	8.1	10.8	16.2
30	0.68	4.5	5.8	6.8	9.0	13.5
35	0.58	3.9	4.9	5.8	7.7	11.6
40	0.51	3.4	4.3	5.1	6.8	10.1
45	0.45	3.0	3.8	4.5	6.0	9.0

Using the above table, you can readily determine how to mix a specific concentration of fertilizer stock solution for injection through most commercially available proportioners. Consider the following example:

You have a 1:200 proportioner and would like to apply 100 ppm N to juniper liners using 20-20-20 soluble fertilizer. How much 20-20-20 should you dissolve per gallon of stock solution in order to have 100 ppm N coming out of the proportioner delivery hose end? (Keep in mind that 20-20-20 contains 20% N, 20% P_2O_5, and 20% K_2O.)

Step 1. In the column labeled "% N,P_2O_5 or K_2O in fertilizer," find 20 since there is 20% N in the fertilizer you're using.

Step 2. Go straight across to the column labeled "1:200" under the heading "Proportioner dilution ratio." Your finger should be on 13.6 ounces of soluble 20-20-20 in 1 gallon of water that when put on through the 1:200 proportioner would yield 100 ppm N. This solution would also contain 100 ppm P_2O_5 and 100 ppm K_2O. In this example, if you want to apply 200 ppm N using the same fertilizer and proportioner, double the fertilizer ounces required per gallon to dissolve 27.2 ounces of fertilizer per gallon of stock solution.

Volume: Dry Measure Equivalents

Unit	Equivalent values
Acre foot	43,560 cubic feet or 1,163$^{1}/_{3}$ cubic yards
Acre inch	3,630 cubic feet or 134$^{1}/_{2}$ cubic yards
Barrel	7,056 cubic inches or 105 quarts dry or 4 cubic feet or 3$^{1}/_{4}$ bushels
Board foot	144 cubic inches (12 inches wide x 12 inches long x 1 inch thick)
Bushel	2,150 cubic inches or 4 pecks or 1$^{1}/_{4}$ cubic feet
Cord	128 cubic feet (4 feet wide x 4 feet high x 8 feet long)
Cubic foot	1,728 cubic inches
Cubic inch	.00057 cubic feet
Cubic yard	21.6 bushels or 27 cubic feet
Cup	48 teaspoons or 16 tablespoons or 2 cups dry or $^{1}/_{2}$ quart dry
Quart	67.2 cubic inches or 4 cups dry or 2 pints dry
Tablespoon	3 teaspoons or .333 ounces
Teaspoon	$^{1}/_{3}$ tablespoons or .111 ounce

Volume: Liquid Measure Equivalents

Unit	Equivalent values
Acre foot	325,151 gallons or 43,560 cubic feet
Acre inch	27,154 gallons or 3,630 cubic feet
Barrel	126 quarts or 31$^{1}/_{2}$ gallons
Cubic inch	0.554 fluid ounce or 0.034 pint
Cubic foot	953 fluid ounces or 59.84 pints or 29.9 quarts or 7.4 gallons
Cup	16 tablespoons or 8 fluid ounces or 2 gills or $^{1}/_{2}$ pint
Gallon	231 cubic inches or 128 fluid ounces or 8 pints or 4 quarts
Gill	4 fluid ounces or $^{1}/_{2}$ cup or $^{1}/_{4}$ pint
Hogshead	63 gallons or 2 barrels
Keg	62$^{1}/_{2}$ quarts or 15$^{2}/_{3}$ gallons
Ounce	6 teaspoons or 2 tablespoons or 1.8 cubic inches
Pint	28.87 cubic inches or 16 fluid ounces or 4 gills or 2 cups
Quart	57.75 cubic inches or 32 fluid ounces or 8 gills or 2 pints
Tablespoon	3 teaspoons or $^{1}/_{2}$ fluid ounce
Teaspoon	50 to 60 drops or $^{1}/_{3}$ tablespoon or 0.17 fluid ounce

Temperature Conversions

The numbers in the center columns labeled "Reading" refer to the temperature that may be in either degrees Fahrenheit or degrees Celsius. If you want to convert from Celsius to Fahrenheit, the equivalent temperature in Fahrenheit will be on the corresponding line in the right hand column. If you want to convert from Fahrenheit to Celsius, the equivalent temperature will be on the corresponding line in the left hand column.

Example 1. To convert 20 degrees Fahrenheit to degrees Celsius, find 20 in the Reading column. The corresponding number in the Celsius column, -6.7, is the temperature in degrees Celsius.

Example 2. To convert 20 degrees Celsius to degrees Fahrenheit, find 20 in the Reading column. The corresponding number in the Fahrenheit column, 68.0, is the temperature in degrees Fahrenheit.

For temperatures not found in the table below, the following formulas can be used:

Celsius to Fahrenheit = (degrees C x 9/5) + 32 = degrees Fahrenheit
Fahrenheit to Celsius = (degrees F - 32) x 5/9 = degrees Celsius

Celsius	Reading	Fahrenheit	Celsius	Reading	Fahrenheit
-17.8	0	32.0	-4.4	24	75.2
-17.2	1	33.8	-3.9	25	77.0
-16.7	2	35.6	-3.3	26	78.8
-16.1	3	37.4	-2.8	27	80.6
-15.6	4	39.2	-2.2	28	82.4
-15.0	5	41.0	-1.7	29	84.2
-14.4	6	42.8	-1.1	30	86.0
-13.9	7	44.6	-0.6	31	87.8
-13.3	8	46.6	0	32	89.6
-12.8	9	48.2	0.6	33	91.4
-12.2	10	50.0	1.1	34	93.2
-11.7	11	51.8	1.7	35	95.0
-11.1	12	53.6	2.2	36	96.8
-10.6	13	55.4	2.8	37	98.6
-10.0	14	57.2	3.3	38	100.4
-9.4	15	59.0	3.9	39	102.2
-8.9	16	60.8	4.4	40	104.0
-8.3	17	62.6	5.0	41	105.8
-7.8	18	64.4	5.6	42	107.6
-7.2	19	66.2	6.1	43	109.4
-6.7	20	68.0	6.7	44	111.2
-6.1	21	69.8	7.2	45	113.0
-5.6	22	71.6	7.8	46	114.8
-5.0	23	73.4	8.3	47	116.6

Celsius	Reading	Fahrenheit	Celsius	Reading	Fahrenheit
8.9	48	118.4	31.7	89	192.2
9.4	49	120.2	32.2	90	194.0
10.0	50	122.0	32.8	91	195.8
10.6	51	123.8	33.3	92	197.6
11.1	52	125.6	33.9	93	199.4
11.7	53	127.4	34.4	94	201.2
12.2	54	129.2	35.0	95	203.0
12.8	55	131.0	35.6	96	204.8
13.3	56	132.8	36.1	97	206.6
13.9	57	134.6	36.7	98	208.4
14.4	58	136.4	37.2	99	210.2
15.0	59	138.2	37.8	100	212.0
15.6	60	140.0	38.3	101	213.8
16.1	61	141.8	38.8	102	215.6
16.7	62	143.6	39.4	103	217.4
17.2	63	145.4	40.0	104	219.2
17.8	64	147.2	40.5	105	221.0
18.3	65	149.0	41.1	106	222.8
18.9	66	150.8	41.6	107	224.6
19.4	67	152.6	42.2	108	226.4
20.0	68	154.4	42.7	109	228.2
20.6	69	156.2	43.3	110	230.0
21.1	70	158.0	43.8	111	231.8
21.7	71	159.8	44.4	112	233.6
22.2	72	161.6	45.0	113	235.4
22.8	73	163.4	45.5	114	237.2
23.3	74	165.2	46.1	115	239.0
23.9	75	167.0	46.6	116	240.8
24.4	76	168.8	47.2	117	242.6
25.0	77	170.6	48.3	118	244.4
25.6	78	172.4	48.3	119	246.2
26.1	79	174.2	48.8	120	248.0
26.7	80	176.0	54.4	130	266.0
27.1	81	177.8	60.0	140	284.0
27.8	82	179.6	65.5	150	302.0
28.3	83	181.4	71.1	160	320.0
28.9	84	183.2	76.6	170	338.0
29.4	85	185.0	82.2	180	356.0
30.0	86	186.8	87.7	190	374.0
30.6	87	188.6	93.3	200	392.0
31.1	88	190.4	98.8	210	410.0

Index of Host Plants, Diseases and Pests

The following is a brief list of some major diseases and insect and mite pests of major floricultural crops. The list is certainly not exhaustive, of either plants or diseases or pests. Not all of the pests and diseases listed here were specifically discussed in the text. New problems are constantly being recorded. Outdoor or saran-produced crops may have different problems than the same plants produced in the greenhouse. Pest and disease problems will be different according to geography and climate. Use this information as a guide to the types of problems likely to be encountered with a plant type. In most cases with insects and mites, more than one aphid, mealybug or scale species can occur on a host plant, so the pest group is listed rather than all of the species known to occur on that plant. In some cases examples are given.

Host	Diseases	Pests
African violet (Saintpaulia)	Botrytis blight Powdery mildew Root and crown rot Root knot nematode	Aphids Mealybugs Tarsonemid mites Springtails Thrips Fungus gnats White grubs
Ageratum	Botrytis blight Root and crown rot Leaf spots	Greenhouse whitefly Mealybugs Spider mites Aphids Cyclamen mite Caterpillars (cabbage looper, corn earworm, tobacco budworm, greenhouse leaf tier) Plant bugs

Aquilegia	Crown rot Leaf spots Rust	Aphids Columbine borer (*Papaipema purpurifascia*) Columbine leaf miner (*Phytomyza acqilegivora*)
Aster	Root rot (Phytophthora) Fusarium wilt Blight (Rhizoctonia) Leaf spots—several Rust Yellows	Aphids Beetles (blister beetle, Japanese beetle) Leafminers (Phytomyza, Agromyza) Leafhoppers Stalk borer Plant bugs
Azalea	Cylindrocladium blight Phytophthora blight Petal blight Leaf gall Leaf spots—several Web blight (Rhizoctonia) Stem canker Phytophthora wilt and root rot	Azalea leaf miner Aphids Black vine weevil Lace bugs Leaf tier Scales Thrips Whiteflies Spider mites Mealybugs Ambrosia beetles Rhododendron borer
Begonia	Bacterial leaf spot and wilt Powdery mildew Root rot	Aphids Mealybugs Whiteflies Thrips Scales Black vine weevil Tarsonemid mites Fungus gnats
Camellia	Canker and dieback Flower blight	Scales Mealybugs Beetles Thrips Aphids Whiteflies Lepidoptera
Carnation (Dianthus)	Alternaria blight Bacterial leaf spot Bacterial leaf wilt Botrytis blight Fairy ring spot (Heterosporium) Fusarium stem rot Fusarium wilt Petal blight Rust Septoria leaf spot Viruses	Aphids Variegated cutworms Other caterpillars (cabbage looper, leafrollers) Thrips Leaf miners (*Liriomyza* spp.) Plant bugs Beetles (Fuller's rose beetle) Fungus gnats Spider mites

Chrysanthemum	Ascochyta blight	Aphids
	Bacterial blight	Spider mites
	Bacterial leaf spot	Variegated cutworm
	Crown rot (Rhizoctonia)	Other caterpillars (cabbage
	Fusarium wilt	looper, corn earworm,
	Leaf spots—several	beet armyworm, European
	Root rot (Pythium)	corn borer)
	Powdery mildew	Beetles (blister beetles,
	Rust	rose chafer)
	Verticillium wilt	Leaf miners
	Viruses	Thrips
		Plant bugs
		Whiteflies
		Mealybugs
		White grubs
		Fungus gnats
Cineraria	Powdery mildew	Aphids
	Botrytis blight	Variegated cutworms
	Root rot	Other caterpillars (cabbage
	Viruses	looper)
		Whiteflies
		Leaf miners
		Thrips
		Mealybugs
		Fungus gnats
		Spider mites
Coleus	Botrytis blight	Whiteflies
	Root rot	Mealybugs
	Viruses	Plant bugs
		Fungus gnats
		Spider mites
		Slugs
Cyclamen	Botrytis blight	Black vine weevil
	Fusarium wilt	Tarsonemid mites
	Bacterial soft rot	Aphids
	Leaf spots and blights	Thrips
	Stunt (Ramularia)	Scales
		White grubs
		Fungus gnats
Dahlia	Bacterial stem rot	Caterpillars (European corn
	Bacterial wilt	borer, stalk borer)
	Botrytis blight	Leafhoppers
	Fusarium and	Aphids
	Verticillium wilt	Plant bugs
	Leaf spots	Thrips
	Powdery mildew	Spider mites
	Stem rot	Cyclamen mites
	Viruses	

Delphinium	Aster yellow Bacterial rot Bacterial leaf spot Diaporthe blight Crown and root rots— several Fusarium wilts Powdery mildew	Tarsonemid mites Aphids Leaf miners Stalk borers Spider mites
Digitalis	Leaf spot Root and stem rot	Aphids Mealybugs Beetles
Frecsia	Fusarium wilt Bacterial scab Leaf spot	Aphids Thrips Bulb mites
Fuchsia	Rust Crown rot Botrytis blight Pythium root rot Viruses	Whiteflies Mealybugs Aphids Beetles Thrips Spider mites Tarsonemid mites Scales

Geranium	Alternaria leaf spot
	Bacterial leaf spot and blight
	Blackleg (Pythium)
	Boytrytis blight
	Rust
	Viruses
		Aphids
		Whiteflies
		Mealybugs
		Beetles
		Spider mites
		Variegated cutworm
		Other caterpillars (geranium
		  plume moth, cabbage
		  looper, leafroller)
		Plant bugs
		Fungus gnats
		Tarsonemid mites
		Slugs

Gerbera	Phytophthora Powdery mildew	Tarsonemid mites Aphids Whiteflies Leafminers Thrips Spider mites
Gladiolus	Botrytis blights Leaf blight (Stephylium) Leaf spots—several Corn rots—several Neck rot (Rhizoctonia) Scab Viruses	Thrips Plant bugs Aphids (foliar and bulb) Spider mites Caterpillars Mealybugs Bulb mites
Gloxinia	Phytophthora leaf and stem rot Myrotherium rot Viruses	Aphids Thrips Black vine weevil Mealybugs Cyclamen mite

Hibiscus	Leaf spots—several Nectria canker Rust	Whiteflies (especially sweet potato whitefly) Aphids Beetles
Hydrangea	Botrytis blights Powdery mildew Viruses	Beetles (rose chafer) Plant bugs Thrips Aphids Scales Spider mites
Impatiens	Root rot Viruses	Thrips Spider mites Cyclamen mite Aphids Beetles (spotted cucumber beetle) Tarnished plant bug
Iris	Bacterial leaf spot Crown rot Fusarium wilt Viruses	Aphids, including bulb aphids Thrips Iris borer Lesser bulb fly Iris weevil Florida red scale Bulb mites
Kalanchoe	Crown rot (Phytophthora) Powdery mildew	Aphids Mealybugs Tarsonemid mites
Latana	Black mildew Rust Fusarium wilt	Aphids Caterpillars Mealybugs Whiteflies Cyclamen mites
Lily	Botrytis blight Root rot Viruses	Aphids Bulb mites Beetles Stalk borers
Lupine	Leaf blights and spots— several Crown rot (Pellicularia) Powdery mildew Rust	Aphids Plant bugs Fungus gnats Caterpillars Spider mites
Marigold (Tagetes)	Wilt and stem rot (Phytophthora) Botrytis blight Leaf spot	Spider mites Leaf miners Aphids Thrips Plant bugs Leafhoppers Japanese beetles

Orchids	Botrytis blight Leaf spots and blights— several Viruses Bacterial soft rot Rust	Aphids Mealybugs Beetles (weevils and borers) Thrips Plant bugs Scales Orchid fly Slugs
Pansy	Black root rot Pythium root rot Botrytis blight Bacterial leaf spot	Aphids Cutworms Spider mites Mealybugs
Petunia	Alternaria leaf spot Crown rot (Phytophthora) Botrytis blight Pythium root rot Viruses	Flea beetles Caterpillars (woolybear, hornworms) Variegated cutworms Aphids Plant bugs Cyclamen mites
Poinsettia	Bacterial canker Soft rot (Erwinia) Root rot Stem rot Black root rot Scab	Whiteflies Thrips Caterpillars (leafroller) Mealybugs Aphids Scales Fungus gnats Spider mites

Rose	Powdery mildew
	Black spot
	Nematodes
	Verticillium wilt
	Botrytis blight
	Cane cankers
	Rust
	Crown gall
	Downy mildew
	Viruses
		Aphids
		Thrips
		Spider mites
		Caterpillars (cutworms,
		  leafrollers)
		Rose midge
		Beetles (Fuller rose beetle,
		  Japanese beetle, rose chafer)
		White grubs
		Scales
		Whiteflies
		Leafhoppers
		Bristly rose slug
		Plant bugs

Salvia	Root rot Botrytis blight	Whiteflies Aphids Scales

Snapdragon	Powdery mildew	Aphids
	Downy mildew	Thrips
	Rust	Cyclamen mite
	Crown rot	Plant bugs
	Botrytis blight	Mealybugs
	Root rots	Caterpillars (variegated cut-
	Stem rots	worms, cabbage loopers)
		Fungus gnats
		White grubs
		Spider mites
Tulip	Botrytis blight	Aphids
	Bulb rots	Narcissus bulb flies
	Viruses	Bulb mites
Verbena	Bacterial wilt	Aphids
	Botrytis blight	Whiteflies
	Root rot	Thrips
		Caterpillars (woolybear,
		leafroller)
		Fungus gnats
		Spider mites
		Tarsonemid mites
Zinnia	Alternaria blight	Aphids
	Powdery mildew	Whiteflies
	Bacterial leaf spot	Leaf miners
		Plant bugs
		Beetles
		Spider mites
		Thrips

323

Glossary

Abdomen: The last body region of an insect or mite; the "tail" end.

Acaricide (miticide): An agent that destroys mites and ticks.

Acidic: A chemical condition consisting of more hydrogen ions than hydroxyl ions, measured in pH units. A pH less than 7 is considered acidic.

Active ingredient (a.i. or AI): Chemicals in a product that are responsible for the pesticidal effect.

Acute toxicity: The immediate toxicity of a material; to cause injury or death from a single exposure.

Adjuvant: A material mixed with pesticides in order to improve contact with leaves or improve pest control. Examples include wetting agents, spreaders, emulsifiers, stickers, penetrants, buffers and dispersing agents.

Aerosol: A spray consisting of drops less than 50 microns in diameter; colloidal suspension of solids or liquids in air.

Aestivate: To become dormant or inactive during the summer.

Algae: Microscopic plants without true stems, leaves or roots, but which contain chlorophyll. Algae are found on the growing medium surface, on irrigation matting and in other poorly-drained areas.

Alkaline: A chemical condition consisting of a higher concentration of hydroxyl ions than hydrogen ions, measured in pH units. A pH greater than 7 is considered alkaline, or basic.

Arthropod: An animal having a segmented body, an exoskeleton, and jointed legs.

Bacterium (plural: bacteria): A microscopic, one-celled microorganism that lacks chlorophyll.

Biological control: The use of one living organism to control another.

Blight: A disease symptom that results in the sudden death of leaves.

Blotch: A leaf disease symptom that appears as a large, irregular shaped, yellow or brown area.

Botanical pesticide: A pesticide produced from naturally occurring chemicals found in some plants. Examples are nicotine, pyrethrum and rotenone. Botanical materials may or may not be highly toxic to mammals.

Broad-spectrum insecticide: Nonselective; having about the same toxicity to most insects.

Canker: The infected area of a stem, branch or main root that is often swollen or sunken and discolored.

Capillary action: The upward movement of water through tiny pores, such as in growing media or plant stem.

Carbamate insecticide: One of a class of insecticides derived from carbamic acid.

Carrier: An inert material that serves as a diluent or vehicle for the active ingredient or toxicant.

Cast skin: The old, outer skeleton (exoskeleton) left behind after the insect molts (aphid).

Caterpillar: The wormlike larva of a moth, butterfly or skipper that usually has three pairs of jointed legs on the thorax and two or more pairs of legs on the abdomen.

Caudal: At or near the tail.

Centimeter: A metric unit of length; 0.394 inches.

Chlorosis: Yellowing of normally green plant tissues; a common symptom of insect damage, disease or nutrient deficiency.

Chronic toxicity: The toxicity of a material determined beyond 24 hours and usually after several weeks of exposure.

Compatible chemicals: Chemicals that can be mixed together without losing their effectiveness.

Cocoon: A silken or fibrous case spun by a larva to provide protection during its pupal period.

Coldfogger: A low-volume pesticide applicator that produces mainly aerosol-sized drops by mechanical methods, without heat.

Common pesticide name: A common chemical name given to a pesticide. Many pesticides are known by a number of trade brand names but have only one recognized common name. For example, the common name for Orthene insecticide is acephate.

Conidium (see Spore).

Cornicle: One of a pair of "honey tubes" (siphunculi) that extend from the abdomen of an aphid.

Crawler: The mobile, first-instar nymph of a scale insect or whitefly.

Crown rot: Disease that affects the plant stem at the soil line or crown.

Cultivar: A variety of a cultivated plant that differs from others in the same species and keeps the distinguishing features when reproduced.

Cultural control: The use of resistant cultivars, crop rotation, fertilization levels and irrigation practices as a pest management method.

Disease: Any disturbance to a plant that interferes with its health. Any deviation from the normal, healthy growth or development. Diseases can be caused by fungi, bacteria, viruses or environmental imbalances.

Dispersable granule: A dry formulation of small particle size designed to be mixed with water prior to use.

Dorsal: Top or uppermost; pertaining to the back or upper side.

Dose, dosage: Same as rate. The amount of toxicant given or applied per unit of plant, animal or surface.

Drench: A thorough soaking of growing media using a chemical pesticide and water.

Dry flowable: A type of granular pesticide formulation. It is mixed with water and becomes a sprayable suspension; also known as water-dispersable granules.

Dust: A dry, pesticide formulation mixed with finely-ground talc, clay or powdered nut shells. Dusts are applied dry, never with water.

Economic injury level: The pest level at which additional management practices must be employed to prevent economic losses.

Electrostatic sprayer: A low-volume or ultra-low-volume applicator that applies sprays containing electrically charged spray drops.

Elytra: Thickened, leathery forewings that cover the hind wings; common to beetles and earwigs.

Emulsifiable concentrate: Concentrated pesticide formulation containing organic solvents and emulsifiers used to make an emulsion when mixed with water. An emulsion is not a solution, but a mixture of one liquid dispersed in another.

Environmental Protection Agency (EPA): The federal agency responsible for pesticide rules and regulations and all pesticide registrations.

EPA establishment number: A number assigned to each pesticide production plant by EPA. The number indicates the plant at which the pesticide product was produced and must appear on all labels of that product.

EPA registration number: A number assigned to a pesticide product by EPA when the product is registered by the manufacturer or his designated agent. The number must appear on all labels for a particular product.

Eradicant: Applies to fungicides in which a chemical is used to eliminate a pathogen from its host or environment.

Eradicate: To eliminate a particular pest species from a designated area.

Exclusion: Control of a disease or insect by preventing infected plants, insects or mites from entering an area free of these problems.

Exhaust fan: A large diameter fan used to exhaust air from the greenhouse.

Flowable: A liquid pesticide formulation in which a very finely ground, solid particle is mixed in a liquid carrier.

Fly speck: A tiny spot of excrement left by a fly, such as a shore fly.

Formulation: Referring to pesticides, a toxic chemical combined with other ingredients to create a product that can be used to control a pest. A formulation may contain one or more active ingredients. Examples are dusts, wettable powders, emulsifiable concentrates.

Fumigant: A gaseous pesticide form.

Fumigation: To kill pests by pesticide vapor action.

Fungicide: A chemical that kills, inactivates or inhibits fungi.

Fungistatic: Chemical action that inhibits the fungal spore germination.

Granule: A pesticide formulation containing an active ingredient impregnated on particles of fired clay, walnut shells, corn cobs or other porous material.

Growing medium: Usually, a mixture of several organic and/or inorganic materials in which plants are grown.

Grub: Usually refers to a slow-moving, C-shaped larva of the Coleoptera (beetle) family Scarabaeidae having three pairs of forelegs and a fat, whitish body; sometimes also used to refer to many growing medium-inhabiting larvae of Coleoptera and Hymenoptera.

Hectare (ha): A metric unit of area; approximately 2.5 acres.

High-volume spray (HV): Generally, a "wet" spray applying more than 50 gallons per acre (500 liters per hectare).

Holistic health management: An approach to maintaining plant health that considers all aspects of the plant's environment.

Honeydew: A sticky, sugary liquid excreted by certain insects of the order Homoptera, including aphids, soft scales, mealybugs and whiteflies.

Host: The plant or animal upon which a parasite lives, or which a predator consumes.

Humidity: The amount of moisture in the air, usually given in an amount relative to air temperature: relative humidity.

Hydraulic sprayer: A power sprayer used to apply high volume sprays.

Hygrometer: An instrument used to determine the relative humidity.

Hypha: A thread of fungal cells; many hyphae together form a mycelium or fungal colony.

Inert ingredient: Material added to chemical pesticides that carry and dilute the active ingredients. They generally have no direct pesticidal activity, but rather influence the behavior of the toxic ingredient(s).

Infectious disease: A fungus, bacteria or virus-caused disease that can be transfererred to other plants.

Infestation: As related to insects and mites, the presence of large numbers of an animal pest species where they are likely to cause plant injury and/or annoyance.

Inhalation: Exposure of test animals either to vapor or dust for a predetermined time.

Inhalation toxicity: To be poisonous to man or animals when breathed into the lungs.

Inoculum: A material containing microorganisms that is associated with the spread of the pathogen from plant to plant.

Inoculum potential: The ability of a fungus or bacteria to infect other plants in any given situation.

Insect-growth regulator (IGR): Chemical substance that disrupts the action of insect hormones controlling molting, maturity from pupal stage to adult and others.

Integrated pest management (IPM): Using multiple tactics to control a pest or pathogen.

Insect: A six-legged arthropod that has three distinct body regions (head, thorax, abdomen) as an adult, and often has one or two pairs of wings.

Instar: The life stage of an arthropod between successive molts.

Larva (plural larvae): In insects with complete metamorphosis, the immature form occurring between the egg and pupal stages (e.g. egg, larva, pupa, adult); in mites, the six-legged first instar.

LD 50: A lethal dose for 50 percent of a population of test animals, expressed as milligrams of toxicant per kilogram of body weight (mg./kg). The higher the LD 50, the more chemical it takes to cause the 50 percent mortality, and the safer the chemical.

Leach watering: The process of thoroughly watering plants to dissolve and wash fertilizer nutrients (salts) out of the media.

Life cycle: The development of an insect or mite from the egg (or live birth) to the reproductive stage.

Low-volume spray (LV): A spray, usually with a large portion of the volume consisting of aerosol-sized drops, with a total volume less than 50 gallons per acre (500 liters per ha).

Maggot: Usually refers to fly larvae without distinct heads or legs.

Metamorphosis: Change in form and function during the development of an insect or mite.

Meter: A metric unit of length; 1.094 yards.

Microbial pesticide: Bacteria, fungi or viruses formulated as pesticides and normally applied with conventional application equipment.

Micron: $1/1000$ (0.001) millimeter; commonly used to classify spray drops.

Millimeter: A metric unit of length; 0.034 inches.

Mite: A tiny arthropod, eight-legged as an adult and closely related to ticks.

Molt: The process of replacing the skin with a new skin; shedding.

Mycelium: A mass of hyphal, a fungal colony.

Natural enemies: Often used to describe parasites, predators, nematodes, and pathogenic microorganisms used in biological control.

Nematode: A tiny roundworm that generally cannot be seen without a microscope. Many are parasitic on plants or insects.

Noninfectious disease: A disease that is caused by environmental imbalances.

Nymph: In reference to insects with simple or no metamorphosis, the immature form between egg and adult; in reference to mites and ticks, the eight-legged, immature form.

Oral toxicity: Toxicity of a compound when given by mouth in a single dose. Usually expressed in milligrams of chemical per kilogram of body weight that kills 50 percent of the test animals. The smaller the number, the greater the toxicity.

Organism: A life form in its entirety.

Organochlorine insecticide: One of the many chlorinated insecticides (e.g., DDT, dieldrin, chlordane, BHC, Lindane).

Organophosphate: Class of insecticides (also one or two herbicides and fungicides) derived from phosphoric acid esters.

Oviposition: The process of laying eggs.

pH: A measure of acidity or alkalinity, measured on a scale of 1 to 14. A pH of 7 is neutral, less than 7 is acidic, greater than 7 is basic.

Parasite: Any plant or animal that lives in or on another organism to the detriment of the host.

Pasteurization: A process in which substances are heated to 140 to 160 F for one hour to kill disease-causing and other harmful pathogens and insects, but retain beneficial pathogens. Often used on soil mixes using aerated steam.

Pathogen: Any disease-producing microorganism, such as a fungus, bacterium or virus.

Parthenogenetic: Capable of reproduction without mating (without male fertilization of the eggs).

Pest: An unwanted bacterium, fungus, virus, insect, mite or other living organism.

Pest scouting: An organized program to detect and monitor for the occurrence of pests.

Pesticide: An "economic poison," defined in most state and federal laws as any substance used for controlling, preventing, destroying, repelling, or mitigating any pest. Includes fungicides, herbicides, insecticides, nematicides, rodenticides, desiccants, defoliants and plant growth regulators.

Physical control: The use of screens, traps, barriers, etc., to reduce or prevent pest infestations.

Phytotoxic: Injurious to plants.

Poison control center: Information sources for human poisoning cases, including pesticides, usually located at major hospitals.

Pore space: The space between the particles of the growing medium.

Predator: A natural enemy that preys on and consumes all or part of its host.

Protectant: Fungicide applied to plant surface before pathogen attack to prevent penetration and subsequent infection.

Pseudopupa: A non-feeding stage in thrips development intermediate between the nymph and adult.

Pupa (plural pupae): In insects with complete metamorphosis, the life stage between larva and adult; also, the next to the last developmental stage in thrips, male scales, and whiteflies is often referred to as a pupa.

Rate: Refers to the amount of active ingredient material applied to a unit area regardless of percentage of chemical in the carrier (dilution).

Reentry (intervals): Waiting interval required by federal law between application of certain hazardous pesticides to crops and the entrance of workers into those crops without having to wear protective clothing. May also be referred to as the waiting period.

Registered pesticides: Pesticide products that have been approved by the Environmental Protection Agency for the uses listed on the label.

Resistance (insecticide): Natural or genetic ability of an organism to tolerate the poisonous effects of a toxicant.

Rotary atomizer: A pesticide applicator that disperses pesticide off of a rapidly rotating, notched disc. A fan may be attached to provide addtitional forward momentum to spray drops.

Secondary pest: A pest that usually does little if any damage but can become a serious pest under certain conditions; e.g., when insecticide applications destroy a given insect's predators and parasites.

Selective pesticide: One that, while killing the pest individuals, spares much or most of the other fauna or flora, including beneficial species, either through differential toxic action or through the manner in which the pesticide is used (formulation, dosage, timing, placement, etc.).

Signal word: A required word that appears on every pesticide label to denote the relative toxicity of the product. The signal words are either **Danger-Poison** for highly toxic compounds, **Warning** for moderately toxic or **Caution** for slightly toxic.

Smoke generator: A pesticide combined with an oxidizing agent in a container that is ignited to produce a toxic vapor or tiny particles.

Soil application: Application of pesticide made primarily to soil surface rather than to vegetation.

Soluble powder: A dry powder pesticide formulation that dissolves in water.

Sooty mold: A dark, fungal growth that develops on foliage covered with honeydew.

Spore: A tiny, reproductive body produced by a plant. For fungi and ferns spores are a major means of reproduction and spread. Often called conidia.

Spreader: Ingredient added to spray mixture to improve contact between pesticide and plant surface. Spreaders generally increase the the area that a given volume of liquid will cover on a leaf.

Sticker: Ingredient added to spray or dust to improve its adherence to plants.

Surfactant: Ingredient that aids or enhances the surface-modifying properties of a pesticide formulation (wetting agent, emulsifier or spreader).

Symptom: A visible or otherwise detectable abnormality caused by a disease or disorder. A condition that indicates a health problem.

Systemic: A material such as a bacterium or pesticide that is absorbed and translocated throughout the plant or animal.

Target: The plants, animals, structures, areas or pests to be treated with a pesticide application.

Thermal pulse-jet applicator: A type of low-volume pesticide application equipment in which a pesticide mixed with an oil or water is injected into a very hot air stream, condenses into a fog, and is carried throughout the area to be treated.

Thorax: The middle region of an insect's body; if present, the wings are found here.

Ultra-low-volume (ULV): Sprays, diluted or undiluted, that are applied at 0.5 gallon or less per acre.

Vector: An agent that transmits or spreads a disease-causing pathogen. For example, certain thrips species are vectors of tomato spotted wilt virus.

Wettable powder: Pesticide formulation of toxicant mixed with inert dust and a wetting agent that mixes readily with water and forms a short term suspension (requires tank agitation).

Wetting agent: A compound that causes spray solutions to contact plant surfaces more thoroughly.

Index